AF361967

A Changing Perspective for Treatment of Chronic Kidney Disease

A Changing Perspective for Treatment of Chronic Kidney Disease

Editor

Giacomo Garibotto

MDPI • Basel • Beijing • Wuhan • Barcelona • Belgrade • Manchester • Tokyo • Cluj • Tianjin

Editor
Giacomo Garibotto
University of Genoa
Italy

Editorial Office
MDPI
St. Alban-Anlage 66
4052 Basel, Switzerland

This is a reprint of articles from the Special Issue published online in the open access journal *Journal of Clinical Medicine* (ISSN 2077-0383) (available at: https://www.mdpi.com/journal/jcm/special_issues/Treatment_CKD).

For citation purposes, cite each article independently as indicated on the article page online and as indicated below:

LastName, A.A.; LastName, B.B.; LastName, C.C. Article Title. *Journal Name* **Year**, *Volume Number*, Page Range.

ISBN 978-3-0365-2794-9 (Hbk)
ISBN 978-3-0365-2795-6 (PDF)

Cover image courtesy of Giacomo Garibotto

Contents

Journal of
Clinical Medicine

MDPI

Editorial

A Changing Perspective for Treatment of Chronic Kidney Disease

Giacomo Garibotto

Department of Internal Medicine, University of Genoa, 16128 Genova, Italy; gari@unige.it; Tel.: +39-(0)103538989

Keywords: chronic kidney disease; CKD–MBD; hyperkalemia; vascular calcification

Citation: Garibotto, G. A Changing Perspective for Treatment of Chronic Kidney Disease. *J. Clin. Med.* **2021**, *10*, 3840. https://doi.org/10.3390/jcm10173840

Received: 26 July 2021
Accepted: 6 August 2021
Published: 27 August 2021

Publisher's Note: MDPI stays neutral with regard to jurisdictional claims in published maps and institutional affiliations.

Chronic kidney disease (CKD) is now an enormous worldwide health problem. The incidence and prevalence of CKD is high and still increasing, mainly owing to obesity and diabetes, and is associated with growing morbidity and mortality [1]. Overall CKD mortality has increased by ˜32% over the last 10 years, making it one of the fastest-rising major causes of death, together with diabetes and dementia [2]. CKD is the 12th most common primary cause of death, accounting for about 1 million deaths per year worldwide [1]. CKD and end-stage renal disease (ESRD) are characterized by the progressive development of a series of complications, such as hypertension, left-ventricular hypertrophy [LVH] anemia, hyperkalemia, hypervolemia, hyperphosphatemia with mineral and bone disorders (CKD–MBD), metabolic acidosis, hyperuricemia and wasting; all of these complications have been shown to be associated with adverse outcomes, and can contribute either individually or in association to the cardiovascular morbidity and mortality observed in CKD.

Current management of CKD includes blood pressure control, treatment of albuminuria with angiotensin-converting enzyme inhibitors or angiotensin II receptor blockers, nutritional intervention, avoidance of potential nephrotoxins and obesity, drug dosing adjustments, and cardiovascular risk reduction. Recent progress in our understanding of CKD pathophysiology together with the development of novel therapeutic agents has led to a renewed attention on the treatment of CKD and its associated metabolic complications which are now amenable to intervention. It is this point that led to the creation of this relevant Special Issue. I am happy to know that scientists from several parts of the world responded with enthusiasm to the opportunity, sending papers which involve genetics, pathophysiology and epidemiology together with the novel therapeutical approaches to CKD.

Genome-wide association studies have identified hundreds of loci where genetic variants are associated with CKD; however, more than 90% of these variants are in non-coding regions of the genome and how they cause disease is still unclear. Lysosomal beta-D-mannosidase (*MANBA*) is an exoglycosidase involved in the sequential degradation of the N-glycosylproteins glycans. Recently, the *MANBA* gene was proposed as a kidney disease severity gene. In this issue, Hye-Rim Kim et al. [3] evaluated, by integrating CKD-related variants and kidney expression quantitative trait loci (eQTL) data, the effects of *MANBA* gene variants on CKD and kidney function-related traits in the Korean Genome and Epidemiology Study (KoGES) cohort. Their study observed 20 single nucleotide polymorphisms (SNPs) that showed a statistically significant association with CKD and kidney function-related traits among 229 SNPs of the MANBA gene. In addition, rs4496586, which had the highest significance for CKD, was associated with MANBA gene expression in renal tubules and glomeruli. In conclusion, this study strongly suggests that *MANBA* gene variants are associated with CKD and kidney function-related traits.

Even if obesity and diabetes largely contribute to the CKD epidemics, their interaction with kidney toxins is incompletely understood. Hagedoorn et al. [4] investigated if lifestyle-related exposures (diet and smoking) contribute to blood cadmium and lead concentrations, as well as whether they are associated with the prevalence of diabetic kidney disease (DKD). In a cross-sectional analysis of a cohort of 231 patients with type 2 diabetes included in the DIAbetes and LifEstyle Cohort Twente (DIALECT-1), median cadmium and lead blood concentrations were below acute toxicity values [2.94 nmol/L for cadmium and 0.07 μmol/L for lead, respectively]. However, every doubling of lead concentration was associated with a 1.75 (95% confidence interval [CI]: 1.11–2.74) times higher risk for albuminuria. In addition, both cadmium and lead were associated with an increased risk for reduced creatinine clearance. The association between cadmium and lead and the prevalence of DKD suggests that they might contribute to the development of this major diabetes complication.

It is still challenging to predict acute and chronic kidney injury after kidney surgery. This is mainly due to a scarcity in sensitive and specific biomarkers for predicting CKD progression early. The development of ESRD may take several years, therefore surrogate endpoints such as albuminuria and serum creatinine have been increasingly used in trials over hard endpoints to predict CKD progression. Acute kidney injury (AKI) is a common complication after nephrectomy in renal cell cancer, with radical nephrectomy being associated with a higher risk than partial nephrectomy. Several predictive factors for CKD following radical nephrectomy (RN) or partial nephrectomy (PN) have been identified. However, the association between AKI and long-term renal function after radical nephrectomy has not been evaluated fully. Won Ho Kim et al. [5], in a retrospective study of 558 cases of radical nephrectomy (median follow-up of 35 months), observed that AKI occurred in 43.2% of cases, and, more importantly, CKD3a developed in 40.5% of patients. The incidence of new-onset CKD was significantly higher in patients with AKI than those without at all follow-up time points after surgery. In conclusion, their analysis demonstrated a strong association between AKI after radical nephrectomy and long-term renal functional deterioration.

Preservation of kidney function can improve outcomes and can be achieved through non-pharmacological strategies and CKD-targeted pharmacological interventions. Gout as well as asymptomatic hyperuricemia have been associated with several traditional cardiovascular risk factors both in adults and children with CKD [6]. In vitro studies and animal models support a role for uric acid mediating both hemodynamic and tissue toxicity leading to glomerular and tubule-interstitial damage, respectively. Nevertheless, two recent well-designed trials failed to show any benefit of allopurinol treatment on renal outcomes, casting doubts on the expectations of renal protection by the use of urate lowering treatment. In addition, a trend for increased mortality was observed in the interventional arms. Russo et al. [7] critically reviewed results from all available randomized controlled trials comparing a urate-lowering agent with placebo or no study medication for at least 12 months. According to the authors, the analysis of the literature does seem to leave it open to the possibility of demonstrating the beneficial effect of urate-lowering agents in future trials. In consideration that both vascular and kidney damages induced by uric acid cannot regress once they have been established, patients with better-preserved renal function and children might benefit more from an early treatment. Adequately powered, randomized, placebo-controlled trials with appropriate selection criteria are needed to determine whether specific patient groups could benefit from urate-lowering agents.

Early identification of the risk factors for CKD and its progression is critical for prevention of kidney damage and adverse outcomes. Cheng-Sheng Yu et al. [8] applied the clustering heatmap and random forest methods to provide an interactive visualization of patients with different CKD stages in a retrospective cohort study. They observed that an index of body composition (waist circumference) and a few biochemical parameters (uric acid, blood urea nitrogen, serum glutamic oxaloacetic transaminase, and HbA1c) were significantly associated with CKD. In their analysis, CKD was associated with obesity, hyperglycemia, and liver function. Interestingly, hypertension and HbA1c levels were associated in the same cluster with a similar pattern. Despite the study limitations (inherent to its retrospective, cross-sectional cohort approach), their data suggest that the clustering heatmap may provide a new predictive model for a high risk of rapid CKD progression.

Continuous growth in the incidence of DKD is the main driver of CKD burden [1,2]. DKD is a major cause of morbidity and mortality in diabetes [9]. Despite advances in the nephroprotective treatment of diabetes, DKD remains the most common complication, driving the need for renal replacement therapies [RRT] worldwide, and its incidence is increasing. Until recently, prevention of DKD progression was based around strict blood pressure [BP] control, using renin–angiotensin system blockers that simultaneously reduce BP and proteinuria, adequate glycemic control and control of cardiovascular risk factors. New drugs which modify intrarenal haemodynamics (such as renin–angiotensin-aldosterone pathway modulators and SGLT2 inhibitors) can preserve the kidney from damage by decreasing intraglomerular pressure independently of blood pressure and glucose control, whereas other novel agents (such as mineralocorticoid receptor antagonists) might offer kidney protection through their antifibrotic mechanisms in DKD. In this issue, Gorriz et al. [10] review the potential of Glucagon-like peptide-1 Receptor Agonist (GLP-1RA) for adequate glycemic control in multiple stages of DKD without increasing risk of hypoglycemia; in addition, GLP-1RA may prevent the onset of macroalbuminuria and slow the decline of glomerular filtration rate (GFR) in diabetic patients, also offering additional benefit in weight reduction, cardiovascular and other kidney outcomes. Trials to assess the impact of GLP-1RA treatments on primary kidney endpoints in DKD are ongoing and some of them will be soon available.

Cardiovascular [CV] disease is a leading cause of morbidity and mortality in patients with CKD and in those on hemodialysis [HD] [11] and is strongly associated with atherosclerosis and vascular calcification [VC]. Patients with CKD have a higher prevalence of vascular calcifications as renal function declines, which will result in increased mortality. Serum calciprotein particles [CPPs] are colloidal nanoparticles that have a prominent role in the initiation and progression of VC. In this issue, Silaghi et al. [12] reviewed the usefulness of the T_{50} test, a novel test that measures the conversion of primary to secondary CPPs, indicating the tendency of serum to calcify, in the assessment of VC. They also made a comprehensive review of the regulation of serum CPP levels, and explored the effects of CPPs and calcification propensity on outcomes. In addition, new topics were raised regarding possible clinical uses of T_{50} in the assessment of VC, particularly in patients with CKD, including possible opportunities in VC management.

Hyperphosphatemia is a common complication of CKD. Even if severe hyperphosphatemia is clinically asymptomatic, it is associated with morbidity and poor outcome. Hyperphosphatemia is an emerging cardiovascular risk factor in CKD, by contributing to vascular calcification. However, mechanisms by which hyperphosphatemia is associated to CV complications are not clearly understood. Abbasian et al. [13] studied the effects of phosphate [Pi] on the release of pro-coagulant activity from endothelial microvesicles [MV] in male Sprague–Dawley rats with experimental CKD; rodents were randomly allocated to receiving high [1.2%] or low [0.2%] dietary phosphorus; and sham-operated controls receiving high [1.2%] phosphorus. After 14 days, as compared to sham controls, high-phosphorus CKD rats presented elevated total plasma MVs, expressed higher CD144 (a major component of endothelial adherens junctions which is expressed by endothelial cells during development), and enhanced procoagulant activity. The results observed in the rat model by Abbasian et al. [13] show that hyperphosphatemia may induce an increase in circulating pro-coagulant MVs, suggesting an important link between elevated circulating phosphate and thrombotic risk in CKD.

Both in CKD and HD patients, volume and pressure overload ultimately lead to LVH, a significant predictor of increased CV events [14]. Although many studies have used pre-HD blood pressure (BP) to determine optimal BP levels, the optimal timing and measurement techniques of blood pressure in HD patients are not well established. In a prospective observational study, Hiroaki et al. [15] aimed at identifying the ideal timing and setting for measuring BP; they observed an association between increased CV events and

LVMI > 156 g/m^2 in patients with diabetes mellitus, after performing multivariate regression analysis. In addition, they found that pre-HDBP at the start of the week, post-HDBP at the end of the week, and weekly averaged BP (WABP) were independently associated with LVMI on univariate regression analysis of follow-up. Multiple BP measurements taken before, during and after dialysis were confirmed to be the most accurate assessment format, a finding which suggests that multiple measurements of BP should be performed and then averaged in this clinical setting.

Hyperkalemia is a very common CKD complication, which accounts for a large number of urgent visits in emergency departments, as well as high mortality. Hyperkalemia is commonly observed in patients with chronic heart failure [CHF], in CKD stage 3–5, and in patients with diabetes mellitus. Among CKD patients, those requiring dialysis represent a group at particularly high risk of hyperkalemia. For a long time, the only therapeutic option for increasing fecal K$^+$ excretion has been represented by sodium polystyrene sulfonate, a cation-exchanging resin. Recently, new drugs able to promote gastrointestinal potassium elimination, namely patiromer and sodium zirconium cyclosilicate, have been developed and studied in large trials, proving their efficacy and safety in different clinical contexts. In this review, Esposito et al. [16] have reviewed the pathophysiology of hyperkalemia, focusing on the mechanisms of action and the clinical data of patiromer and sodium zirconium cyclosilicate, considering that these new treatments may represent a chance to improve the management of both acute and chronic hyperkalemia.

Assessing cognitive, nutritional and functional status in elderly subjects with CKD is emerging as a new tool to stratify the risk to develop ESRD and death [17]. In addition, the prognostic evaluation of older adults with CKD is crucial to identify the most appropriate clinical decision-making process for patients and their families. A multidimensional assessment (MPI) may represent an important aspect in predicting short- and long-term all-cause mortality in elderly patients with CKD [18]. In particular, in CKD patients, the MPI was shown to be more accurate in predicting mortality when compared to the eGFR alone [18]. In this issue, Lai et al. [19] longitudinally studied the associations in MPI, both the hospitalization and mortality in clinically stable CKD (n = 105) patients on conservative therapy (eGFR $\leq$ 60 mL/min, stage 3–5 KDOQI), or renal replacement therapy (HD = 32 pts or PD = 36 pts), for at least 3 months. A total of 173 patients, with a median age of 76 years, was studied. The median duration of all the hospitalizations was 6 days and the number of deaths was 33. MPI significantly correlated with days and number of hospitalizations per year. According to the findings, MPI was associated with outcomes in patients with renal disease, suggesting that a multidimensional evaluation should be implemented in this clinical context.

Many of the important issues dealing on prevention and treatment of CKD are addressed in this volume. We thank the writers, the large number of investigators and the MDPI staff for their leadership in producing this Special Issue. We hope the readers will enjoy and benefit from new insight.

Funding: This research received no external funding.

Institutional Review Board Statement: Not applicable.

Informed Consent Statement: Not applicable.

Data Availability Statement: Not applicable.

Conflicts of Interest: The author declares no conflict of interest.

References

1. Brendon, L.N.; Chadban, S.G.; Demaio, A.L.; Johnson, D.W.; Perkovic, V. Chronic kidney disease and the global NCDs agenda. *BMC Glob. Health* **2017**, *2*, e000380.

2. Wang, H.; Naghavi, M.; Allen, C.; Barber, R.M.; Bhutta, Z.A.; Carter, A.; Casey, D.C.; Charlson, F.J.; Chen, A.Z.; Coates, M.; et al. GBD 2015 Mortality and Causes of Death Collaborators. Global, regional, and national life expectancy, all-cause mortality, and cause-specific mortality for 249 causes of death, 1980–2015: A systematic analysis for the Global Burden of Disease Study 2015. *Lancet* **2016**, *388*, 1459–1544. [CrossRef]

3. Kim, H.R.; Jin, H.S.; Eom, Y.B. Association between *MANBA* Gene Variants and Chronic Kidney Disease in a Korean Population. *J. Clin. Med.* **2021**, *10*, 2255. [CrossRef] [PubMed]

4. Hagedoorn, C.M.; Gant, S.; Huizen, V.; Maatman, R.G.H.; Navis, G.; Bakker, S.J.L.; Laverman, G.D. Lifestyle-Related Exposure to Cadmium and Lead is Associated with Diabetic Kidney Disease. *J. Clin. Med.* **2020**, *9*, 2432. [CrossRef] [PubMed]

5. Kim, W.O.; Shin, K.W.; Ji, S.H.; Jang, Y.E.; Lee, I.J.; Jeong, C.W.; Kwak, C.; Lim, Y.J. Robust Association between Acute Kidney Injury after Radical Nephrectomy and Long-term Renal Function. *J. Clin. Med.* **2020**, *9*, 619. [CrossRef] [PubMed]

6. Viazzi, F.; Rebora, P.; Giussani, M.; Orlando, A.; Stella, A.; Antolini, L.; Valsecchi, M.G.; Pontremoli, R.; Genovesi, S. Increased Serum Uric Acid Levels Blunt the Antihypertensive Efficacy of Lifestyle Modifications in Children at Cardiovascular Risk. *Hypertension* **2016**, *67*, 934–940. [CrossRef] [PubMed]

7. Russo, E.; Verzola, D.; Leoncini, G.; Cappadona, F.; Esposito, P.; Pontremoli, R.; Viazzi, F. Treating Hyperuricemia: The Last Word Hasn't Been Said Yet. *J. Clin. Med.* **2021**, *10*, 819. [CrossRef] [PubMed]

8. Yu, C.S.; Lin, C.H.; Lin, Y.J.; Lin, S.Y.; Wang, S.T.; Wu, J.L.; Tsai, M.H.; Chang, S.S. Clustering Heatmap for Visualizing and Exploring Complex and High-dimensional Data Related to Chronic Kidney Disease. *J. Clin. Med.* **2020**, *9*, 403. [CrossRef] [PubMed]

9. De Cosmo, S.; Viazzi, F.; Pacilli, A.; Giorda, C.; Ceriello, A.; Gentile, S.; Russo, G.; Rossi, M.C.; Nicolucci, A.; Guida, P.; et al. Achievement of therapeutic targets in patients with diabetes and chronic kidney disease: Insights from the Associazione Medici Diabetologi Annals initiative. *Nephrol. Dial. Transplant.* **2015**, *30*, 1526–1533. [CrossRef] [PubMed]

10. Górriz, J.L.; Soler, M.J.; Navarro-González, J.F.; García-Carro, C.; Puchades, M.J.; D'Marco, L.; Castelao, A.M.; Fernández-Fernández, B.; Ortiz, A.; Górriz-Zambrano, C.; et al. GLP-1 Receptor Agonists and Diabetic Kidney Disease: A Call of Attention to Nephrologists. *J. Clin. Med.* **2020**, *9*, 947. [CrossRef] [PubMed]

11. Saran, R.; Robinson, B.; Abbott, K.C.; Bragg-Gresham, J.; Chen, X.; Gipson, D.; Gu, H.; Hirth, R.A.; Hutton, D.; Jin, Y.; et al. US Renal Data System 2019 Annual Data Report: Epidemiology of kidney disease in the United States. *Am. J. Kidney Dis.* **2020**, *75*, A6–A7. [CrossRef] [PubMed]

12. Silaghi, C.N.; Ilyés, T.; Van Ballegooijen, A.J.; Crăciun, A.M. Calciprotein Particles and Serum Calcification Propensity: Hallmarks of Vascular Calcifications in Patients with Chronic Kidney Disease. *J. Clin. Med.* **2020**, *9*, 1287. [CrossRef] [PubMed]

13. Abbasian, N.; Goodall, A.H.; Burton, J.O.; Bursnall, D.; Bevington, A.; Brunskill, N.J. Hyperphosphatemia Drives Procoagulant Microvesicle Generation in the Rat Partial Nephrectomy Model of CKD. *J. Clin. Med.* **2020**, *9*, 3534. [CrossRef] [PubMed]

14. Foley, R.N.; Curtis, B.M.; Randell, E.W.; Parfrey, P.S. Left ventricular hypertrophy in new hemodialysis patients without symptomatic cardiac disease. *Clin. J. Am. Soc. Nephrol.* **2010**, *5*, 805–813. [CrossRef] [PubMed]

15. Io, H.; Nakata, J.; Inoshita, H.; Ishizaka, M.; Tomino, Y.; Suzuki, Y. Relationship among Left Ventricular Hypertrophy, Cardiovascular Events, and Preferred Blood Pressure Measurement Timing in Hemodialysis Patients. *J. Clin. Med.* **2020**, *9*, 3512. [CrossRef] [PubMed]

16. Esposito, P.; Conti, N.E.; Falqui, V.; Cipriani, L.; Picciotto, D.; Costigliolo, F.; Garibotto, G.; Saio, M.; Viazzi, F. New Treatment Options for Hyperkalemia in Patients with Chronic Kidney Disease. *J. Clin. Med.* **2020**, *9*, 2337. [CrossRef] [PubMed]

17. Angleman, S.B.; Santoni, G.; Pilotto, A.; Fratiglioni, L.; Welmer, A.K. Multidimensional Prognostic Index in Association with Future Mortality and Number of Hospital Days in a Population-Based Sample of Older Adults: Results of the EU Funded MPI_AGE Project. MPI_AGE Project Investigators. *PLoS ONE* **2015**, *10*, e0133789. [CrossRef] [PubMed]

18. Pilotto, A.; Panza, F.; Sancarlo, D.; Paroni, G.; Maggi, S.; Ferrucci, L. Usefulness of the multidimensional prognostic index [MPI] in the management of older patients with chronic kidney disease. *J. Nephrol.* **2012**, *25*, 79–84. [CrossRef] [PubMed]

19. Lai, S.; Amabile, M.I.; Mazzaferro, S.; Imbimbo, G.; Mitterhofer, A.P.; Galani, A.; Aucella, F.; Brunori, G.; Menè, P.; Molfino, A. Association between Multidimensional Prognostic Index and Hospitalization and Mortality among Older Adults with Chronic Kidney Disease on Conservative or on Replacement Therapy. *J. Clin. Med.* **2020**, *9*, 3965. [CrossRef] [PubMed]

Journal of *Clinical Medicine*

Article

Clustering Heatmap for Visualizing and Exploring Complex and High-dimensional Data Related to Chronic Kidney Disease

Cheng-Sheng Yu [1,2,†], Chang-Hsien Lin [1,2,†], Yu-Jiun Lin [1,2], Shiyng-Yu Lin [1,2], Sen-Te Wang [1,2], Jenny L Wu [1,2], Ming-Hui Tsai [3] and Shy-Shin Chang [1,2,*]

[1] Department of Family Medicine, Taipei Medical University Hospital, Taipei 110, Taiwan; molytrigger@gmail.com (C.-S.Y.); 862077@h.tmu.edu.tw (C.-H.L.); b101096105@hotmail.com (Y.-J.L.); daleslin@gmail.com (S.-Y.L.); wangader@gmail.com (S.-T.W.); jenny31325et@gmail.com (J.L.W.)

[2] Department of Family Medicine, School of Medicine, College of Medicine, Taipei Medical University, Taipei 110, Taiwan

[3] Preventive health center, Taipei Medical University Hospital, Taipei 110, Taiwan; 916048@h.tmu.edu.tw

* Correspondence: sschang0529@gmail.com; Tel.: +886-0227-372-18 (ext. 3966)

† Equal contribution authors.

Received: 22 January 2020; Accepted: 31 January 2020; Published: 2 February 2020

Abstract: Background: Preventive medicine and primary health care are essential for patients with chronic kidney disease (CKD) because the symptoms of CKD may not appear until the renal function is severely compromised. Early identification of the risk factors of CKD is critical for preventing kidney damage and adverse outcomes. Early recognition of rapid progression to advanced CKD in certain high-risk populations is vital. Methods: This is a retrospective cohort study, the population screened and the site where the study has been performed. Multivariate statistical analysis was used to assess the prediction of CKD as many potential risk factors are involved. The clustering heatmap and random forest provides an interactive visualization for the classification of patients with different CKD stages. Results: uric acid, blood urea nitrogen, waist circumference, serum glutamic oxaloacetic transaminase, and hemoglobin A1c (HbA1c) were significantly associated with CKD. CKD was highly associated with obesity, hyperglycemia, and liver function. Hypertension and HbA1c were in the same cluster with a similar pattern, whereas high-density lipoprotein cholesterol had an opposite pattern, which was also verified using heatmap. Early staged CKD patients who are grouped into the same cluster as advanced staged CKD patients could be at high risk for rapid decline of kidney function and should be closely monitored. Conclusions: The clustering heatmap provided a new predictive model of health care management for patients at high risk of rapid CKD progression. This model could help physicians make an accurate diagnosis of this progressive and complex disease.

Keywords: heatmap; clustering; multivariate statistical analysis; chronic kidney disease; risk factors

---

1. Introduction

Chronic kidney disease (CKD) poses a substantial challenge to global health policy because of its health and economic burden [1]. Primary health care is vital for CKD because of the high global prevalence (11%–13%) of underdiagnosed and undertreated patients with stage 3 CKD [2]. In Taiwan, the prevalence of CKD is as high as 11.93% (approximately 2.74 million patients), resulting in a potentially high mortality rate because of cardiovascular diseases and accounting for tremendous health care expenditures [3]. Therefore, early identification of the risk factors for CKD is critical for preventing kidney damage and adverse outcomes.

Identification and staging of CKD rely on the measurement of glomerular filtration rate (GFR) and albuminuria. The calculation of actual GFR by measurement of external filtration markers is cumbersome and impractical. Moreover, clinical studies have found that external GFR markers are not reliable predictors of risk of advanced CKD because of the inconsistency in their levels at different stages of CKD [4–8].

Recently, machine learning techniques have showed success in prediction and diagnosis of numerous critical diseases. In fact, one study found that CKD can be diagnosed with an accuracy of 0.98, using supervised machine learning technique [9]. However, to the best of our knowledge, there is still no attempt to classify and visualize CKD staging via machine learning technique. Given that GFR measurement can be cumbersome and inconsistent, we aimed to use unsupervised learning technique, hierarchical clustering with heatmap visualization to classify the CKD staging of patients. The results of our study may identify patients who are categorized as having early stage CKD but have physiological parameters similar to those with advanced-stage CKD. Thus, clinicians could provide active health intervention to this group to minimize their risk of advanced renal disease.

2. Experimental Section

2.1. Study Design

This single-center, retrospective cohort study assessed the ability of the clustering heatmap to classify and predict the risk of CKD in participants who underwent a self-paid health examination at the Health Management Center (HMC) of Taipei Medical University Hospital (TMUH).

2.2. Setting

The study was conducted at TMUH, and patients' electronic medical records were reviewed. TMUH is a private, 800-bed teaching hospital in Taiwan. HMC of the TMUH receives 50–60 visits each day. The Institutional Review Board of TMUH approved this study (TMU-JIRB No.: N201903080), which was conducted in accordance with the original and amended Declaration of Helsinki. IRB approval was obtained for this project before starting data collection and informed consent was waived by the IRB owing to the retrospective nature of the study.

2.3. Inclusion and Exclusion Criteria

The population of interest was participants who underwent a self-paid health examination at the HMC of TMUH. To be included in the analysis, the self-paid health examination participants must have undergone a self-paid health examination comprising of an abdominal transient elastography inspection using FibroScan 502 Touch (Echosens, Paris, France) and urine test from March 2015 to December 2019. Exclusion criteria included, participants the age of participants ≤ 18 years old, and participants that do not have follow-up urine test, within 3 months when the first urine test is found to be abnormal.

2.4. Data

All participants undertook the regular processes of HMC. The participants were interviewed by well-trained personnel, who verified the correctness of the participants' self-completed questionnaire on demographics, existing medical conditions, and use of medication. In addition, the personnel confirmed adherence to health examination prerequisites (e.g., overnight fasting for at least 8 h) for the package chosen by the participants. Those found to have not followed the prerequisites were suggested to book another appointment. Then, anthropometrics (weight, height, waist circumference, and arterial pressures) were measured. The instruments were regularly calibrated per the manufacturer's specifications. The samples of blood, urine, and specimens required per the chosen package were collected for laboratory tests. Regular laboratory test items included alpha-fetoprotein, hemoglobin A1c (HbA1c), serum glutamic oxaloacetic transaminase (GOT), serum glutamic-pyruvic transaminase (GPT), uric acid (UA), creatinine, blood urea nitrogen (BUN), total protein, albumin, globulin, albumin/globulin ratio (A/G), total bilirubin, direct bilirubin, alkaline phosphatase, gamma-glutamyl transpeptidase (r-GT), total cholesterol, low-density lipoprotein cholesterol (LDL), high-density lipoprotein cholesterol (HDL), LDL/HDL ratio, total cholesterol/HDL ratio, triglycerides, fasting blood sugar, and thyroid-stimulating hormone (TSH). Urine specimens were obtained in the morning and scheduled to avoid menstrual periods and check proteinuria, hematuria, red blood cell cast, white blood cell cast, and other urine sediment abnormalities. A repeated check on urine specimens was performed within 3 months, if an abnormal urine test is obtained. For each FibroScan (502 Touch; Echosens, Paris, France) inspection, two scores were reported: controlled attenuation parameter (CAP) score and liver stiffness parameter (E score) [10–13].

2.5. Definitions of Measurement Cutoffs and Calculations

Body mass index (BMI) categories were defined according to the ranges established for Asian populations by the Ministry of Health and Welfare of Taiwan as follows: obesity, $\geq$27 kg/m^2; overweight, 24–26.9 kg/m^2; and normal weight, <23.9 kg/m^2 [14]. The cutoff for waist circumference for abdominal obesity was $\geq$90 cm for men and $\geq$80 cm for women, using the Asian-specific cutoff points established by the International Diabetes Federation [15]. Estimated GFR (eGFR) was calculated using equations for the Modification of Diet in Renal Disease (MDRD) formula for Chinese patients as follows: 175 $\times$ (Scr) $^{-1.234}\times$ (age) $^{-0.179}$ $\times$ 0.79 (if female) [16]. In this study, CKD was classified into five stages using Kidney Outcomes Quality Initiative (KDOQI) [17] guidelines using thresholds of eGFR within the CKD range and/or evidence of structural renal changes, e.g., proteinuria. In our study, we used equations for the Modification of Diet in Renal Disease for Chinese patients with CKD [16] measured in the following manner: 175 $\times$ (Scr) $^{-1.234}$ $\times$ (Age) $^{-0.179}$ $\times$ 0.79 (if female). The definition of the five stages of CKD was stage-1 (eGFR > 90 mL/min/1.73 m^2 with serial proteinuria positive); stage-2 (eGFR 60–89 mL/min/1.73 m^2 with serial proteinuria positive); stage-3 (eGFR 30–59 mL/min/1.73 m^2; Stage-4 (eGFR 29–15 mL/min/1.73 m^2); and stage-5 (eGFR < 15 mL/min/1.73 m^2). Serial proteinuria was defined as two positive proteinuria in two separate urine tests over a period of 3 months.

2.6. Statistical Analysis

Statistical analysis was conducted using R (version 3.6.1) or SPSS (version 17.0) software. In Table 1, the baseline characteristics of the enrollees were described, and the *p*-value denotes comparison between CKD stage 1 and CKD stage 2–5 participants. Binary variables were presented as a frequency and percentage and were compared using the Chi-squared test or Fisher's exact test. The Mann–Whitney U-test, a nonparametric test, was used to compare the medians of continuous variables. In Table 2, multivariate logistic regression was conducted to assess the significance of clinical data, and *p*-value denotes whether the variable is statistically significant. In addition, variance inflation factor (VIF) was used to check multicollinearity [18,19]. In all analysis, *p*-value < 0.05 was considered statistically significant.

2.6.1. Receiver Operating Characteristic Curve

Receiver operating characteristic (ROC) curves were used to illustrate the diagnostic ability of machine learning classification. Area under the curve (AUC), true-positive rate (also called sensitivity or recall), and false-positive rate (related to specificity) are shown in a graphical plot [20].

Table 1. Descriptive statistics and testing of medical variables in health examination data with chronic kidney disease.

Factors	CKD Stage 1 $n_1 = 1715$ No. (%)	CKD Stage 2 $n_2 = 370$ No. (%)	CKD Stage 3–5 $n_3 = 202$ No. (%)	*p*-Value
Sex				
Female	1017 (59.3%)	96(25.9%)	63(31.2%)	<0.001
Male	698 (40.7%)	274(74.1%)	139(68.8%)	
Hypertension				
Normal	1299 (75.7%)	214(57.8%)	112(55.4%)	<0.001
High	416 (24.3%)	156(42.2%)	90(44.6%)	
		Median (IQR)		
Age, years	42 (35–51)	53 (46–60.75)	63.5 (55.25–73)	<0.001
Albumin, g/dL	4.6 (4.4–4.8)	4.6 (4.5–4.8)	4.3 (3.8–4.5)	<0.001
BMI, kg/m^2	23.4 (21.1–25.99)	24.7 (22.7–27.02)	25.05 (22.15–27.98)	<0.001
WC, cm	80.5 (73.5–88)	86 (80–92)	87.58 (82–95.75)	<0.001
AFP, ng/mL	2.35 (1.62–3.27)	2.62 (1.98–3.62)	2.69 (2.15–4.48)	<0.001
ALKp, IU/L	60 (50–72)	67 (55–79)	79.69 (65.68–96)	<0.001
GOT, IU/L	20 (17–24)	23 (20–28)	23.1 (19–34)	<0.001
GPT, IU/L	18 (13–28)	23 (17–33)	22 (15–35)	<0.001
T_Bilirubin, mg/dL	0.6 (0.4–0.8)	0.6 (0.5–0.8)	0.6 (0.4–0.9)	0.002
γGT, U/L	17 (12–27)	23 (17–34)	34.63 (21–70.90)	<0.001
CAPscore, dB/m	241 (208–281)	260 (227–301.8)	246 (203–296.5)	<0.001
Escore, kPa	4.3 (3.5–5.1)	4.6 (3.7–5.6)	7.9 (4.925–13.525)	<0.001
BUN, mg/dL	12 (10–14.34)	15 (13–18)	23.5 (16.78–38.82)	<0.001
Creatinine, mg/dL	0.7 (0.6–0.8)	1 (1–1.1)	1.7 (1.425–3.9)	<0.001
UA, mg/dL	5.1 (4.3–6.3)	6.3 (5.5–7.3)	6.6 (5.7–7.975)	<0.001
Cholesterol, mg/dL	187 (165–209)	191.5 (165–217)	169 (130.2–195)	<0.001
HbA1C, %	5.4 (5.2–5.6)	5.525 (5.3–5.9)	5.7 (5.3–6.3)	<0.001
HDL, mg/dL	54 (45–65)	49 (41–58)	46 (37–57.9)	<0.001
LDL, mg/dL	121 (102–143)	131 (102.2–152)	107 (80.25–128.75)	<0.001
TSH, μIU/mL	1.81(1.2–2.545)	2.075 (1.465–2.835)	2.009 (1.54–2.565)	<0.001

CKD, chronic kidney disease; BMI, body mass index; WC, waist circumference; AFP, alpha-fetoprotein; ALKp, alkaline phosphatase; GOT, serum glutamic-oxalocetic transaminase; GPT, serum glutamic-pyruvic transaminase; γGT, γ-Glutamyl transpeptidase; BUN, blood urea nitrogen; UA, uric acid; HDL, high-density lipoprotein; LDL, low-density lipoprotein; HbA1C, glycated hemoglobin; TSH, thyroid-stimulating hormone; IU, international units; U, μmol/min; μIU, micro-international units. The upper part includes categorical variables and the others are continuous variables, which are general physiological indices in the first section, hepatic indices and nephritic elements in the second section, and blood lipid and thyroid indices in the final section.

2.6.2. Random Forest

Random decision forest, an ensemble learning method for classification, regression, or other applications based on a decision tree structure, was used to extract the most relevant variables from those biomarkers. The random forest model created multiple decision trees and then combined the output generated by each decision tree [21,22]. This approach removed the bias that a decision tree model might introduce in the system, thus considerably improving the predictive power. All procedures in the random forest model were conducted in the R package "randomForest" [23].

2.6.3. Missing Values

Data with missing values were regulated by an expectation–maximization algorithm, which was an iterative procedure and preserved the relationship with other variables. After deletion of variables with missing values >10% of the sample size, 23 variables were retained [24].

Table 2. Multivariate logistic regression analysis of whole biomarkers related to chronic kidney disease.

Factors	Odds Ratio	95% CI OR	VIF	ΔVIF	*p*-Value
BMI, kg/m^2	0.904	(0.854, 0.958)	3.977	3.975	**0.001** *
WC, cm	1.046	(1.023, 1.069)	4.170	4.169	**<0.001** *
Cholesterol, mg/dL	0.983	(0.973, 0.994)	10.598	Δ	**0.002** *
Hypertension	0.213	(0.654, 1.099)	1.123	1.122	0.213
HbA1C, %	**1.248**	(1.087, 1.434)	1.223	1.178	**0.002** *
HDL, mg/dL	1.005	(0.994, 1.016)	1.945	1.334	0.363
LDL, mg/dL	1.016	(1.005, 1.028)	10.286	1.077	**0.004** *
Albumin, g/dL	0.793	(0.536, 1.174)	1.355	1.364	0.247
ALKp, IU/L	1.001	(0.995, 1.006)	1.202	1.209	0.761
GOT, IU/L	1.030	(1.012, 1.050)	3.832	3.801	**0.001** *
GPT, IU/L	0.977	(0.966, 0.988)	3.847	3.803	**<0.001** *
γGT, U/L	1.002	(0.997, 1.006)	1.452	1.448	0.425
T_Bilirubin, mg/dL	1.088	(0.930, 1.273)	1.182	1.180	0.291
CAPscore, dB/m	1.003	(1.000, 1.006)	1.646	1.647	**0.044** †
Escore, kPa	1.012	(0.985, 1.040)	1.622	1.595	0.393
AFP, ng/mL	1.000	(1.000, 1.000)	1.141	1.129	0.528
BUN, mg/dL	**1.229**	(1.190, 1.270)	1.063	1.062	**<0.001** *
UA, mg/dL	**1.478**	(1.349, 1.619)	1.346	1.333	**<0.001** *
TSH, μIU/mL	1.047	(1.004, 1.091)	1.016	1.015	0.031

* indicates the *p* value < 0.001, † indicates the *p*-value < 0.05. BMI, body mass index; WC, waist circumference; HbA1C, glycated hemoglobin; HDL, high-density lipoprotein; LDL, low-density lipoprotein; ALKp, alkaline phosphatase; GOT, serum glutamic-oxalocetic transaminase; GPT, serum glutamic-pyruvic transaminase; γGT, γ-Glutamyl transpeptidase; AFP, alpha-fetoprotein; BUN, blood urea nitrogen; UA, uric acid; TSH, thyroid-stimulating hormone; IU, international units; U, μmol/min; μIU, micro-international units. Chronic kidney disease is the dependent variable, and all the biomarkers are the independent variables in the logistic analysis. The odds ratio represents the exp(β), which is the exponential of the estimator in logistic regression with 95% confidence interval. In addition, the variance inflation factor (VIF) for each variable is calculated to check multicollinearity. Factors with high odds ratio or significant *p*-value are marked in bold. Factors with high VIF values are shaded. ΔVIF records the variance inflation factor (VIF) after removing the predictor variables with high VIF value.

2.6.4. Multivariate Analysis

Multivariate analysis, which involved observation and analysis of more than one statistical outcome variable simultaneously, was employed as the analysis of large multivariable data sets was a major challenge for life science research. Multivariate analysis has been made much easier with inexpensive, fast computers and powerful analytical software. The application of multivariate analysis included dimensionality reduction, clustering, and variable selection. Here, a heatmap and hierarchical clustering were used [25].

2.6.5. Heatmap and Clustering

A heatmap was used to visualize the pattern of the medical variables generated through classification. The heatmap, a graphical representation of data with the individual values contained in a matrix, was represented as grids of colors plus clustering on both rows and columns. Clustering was applied to group a set of patients according to their health examination data. Participants were divided into different clusters using Ward's hierarchical clustering method, and the patterns of their biomarkers were shown in colors in the center of the heatmap. The heatmap was available from the package "gplot" as an enhanced version or its basic function stats in R [26–31].

3. Results

A total of 2287 participants were enrolled. The whole procedure from data collection to statistical analysis and outcome is shown in Figure 1. After a series of data-cleaning procedures, the participants were stratified into three groups (stage 1, stage 2, and stage 3–5) and compared on Table 1. A higher percentage of female was observed in stage 1 (59.3%) as compared to stage 2 (25.9%) and stage 3–5

(31.2%). As expected, stage 1 CKD participants were generally younger and healthier than stage 2 and stage 3–5 participants.

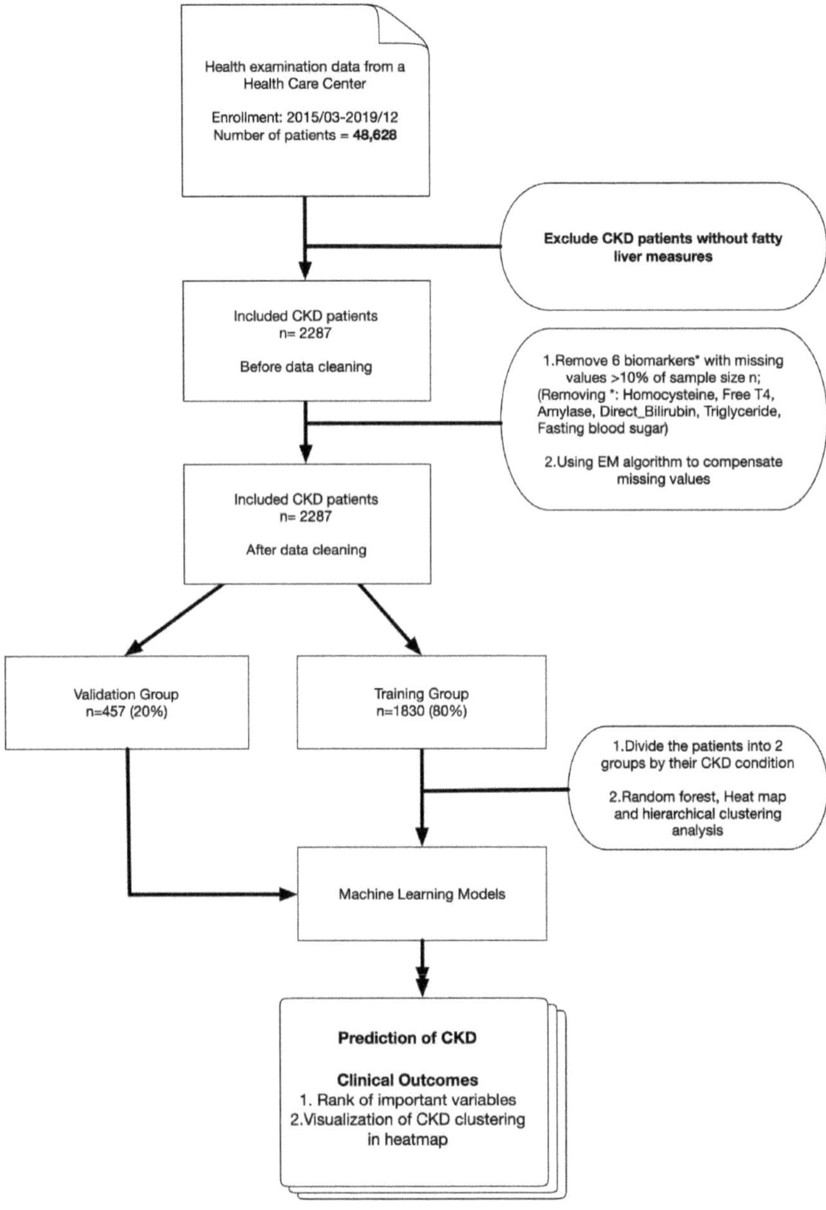

Figure 1. A flowchart of the step-by-step procedure from data collection and preprocessing to statistical analyses.

Variables that had prominent effects on CKD were assessed by multivariate logistic regression (Table 2). Two biomarkers (BUN, odds ratios of 1.229 and UA, odds ratios of 1.478) related to kidney function and one biomarker related to metabolic syndrome (HbA1C, odds ratios of 1.248) were found to have substantial positive association with CKD status. Other metabolic syndrome markers that exhibited significant positive association with CKD status were BMI, and WC. In addition, significant association was observed for three hepatic markers (GOT, GPT and CAPscore). VIF detected multicollinearity between cholesterol and LDL; hence, cholesterol was excluded from subsequent multivariate analyses.

For every variable, the weights of importance measured using random forest are depicted in Figure 2a. The top rank of variables, which had leading scores of mean decrease accuracy in random forest, were similar to the multivariable analysis, Creatinine (87.2%), BUN (24.4%), UA (18.3%), HbA1c (17.2%), Escore (15%), γGT (14.7%), and GPT (14.2%). The AUC of the random forest model using ROC was approximately 0.984 (Figure 2b).

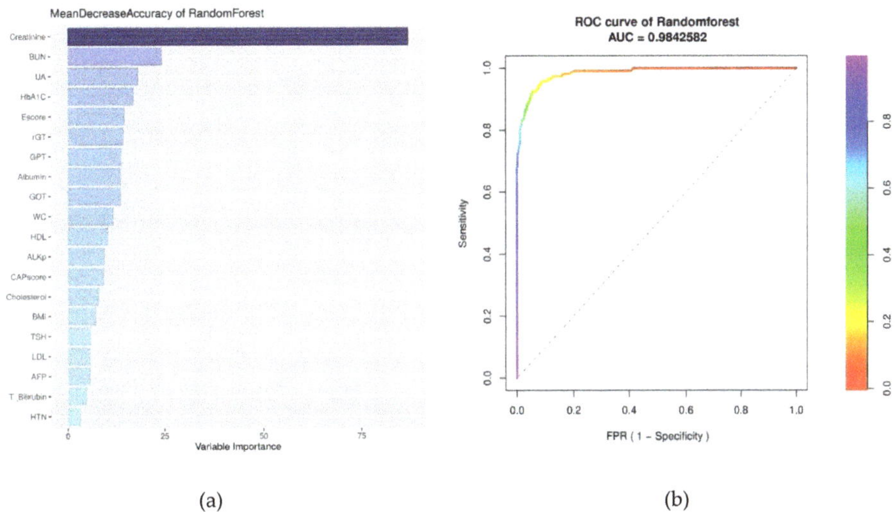

(a)   (b)

Figure 2. (a) Variable importance ordered by the accuracy of mean decrease in random forest. The leading variables obtained by random forest list in **a** with a darker blue; conversely, less prominent variables are indicated in a lighter blue. (b) area under the ROC curve of random forest. The rainbow bar indicates the value of specificity in the false-positive rate.

Finally, multivariate analysis using clustering heatmaps described the classification of CKD stages and related biomarkers (Figure 3). In the row clustering step, most patients with early CKD (stage 1 in dark green; stage 2 in sky blue) were grouped in the lower cluster, whereas most patients with more advanced CKD (stage 3 in pink; stage 4 in red violet; stage 5 in red) were grouped into the center clusters. In addition, every biomarker was clustered into different subgroups on the column side according to their color patterns in the center grids of heatmap. Hepatic biomarkers (blue) were grouped together to the right side with renal biomarkers (light green) because their patterns were more similar than the others. Metabolic (green and salmon) and lipid biomarkers (light blue) were grouped into the left cluster as their patterns were more similar. HDL had an opposite color pattern because higher HDL indicates a healthy state. Heatmaps provided a systematic and clustered visualization of the analyzed data, facilitating monitoring of CKD patients according to their health examination data (Figure 3).

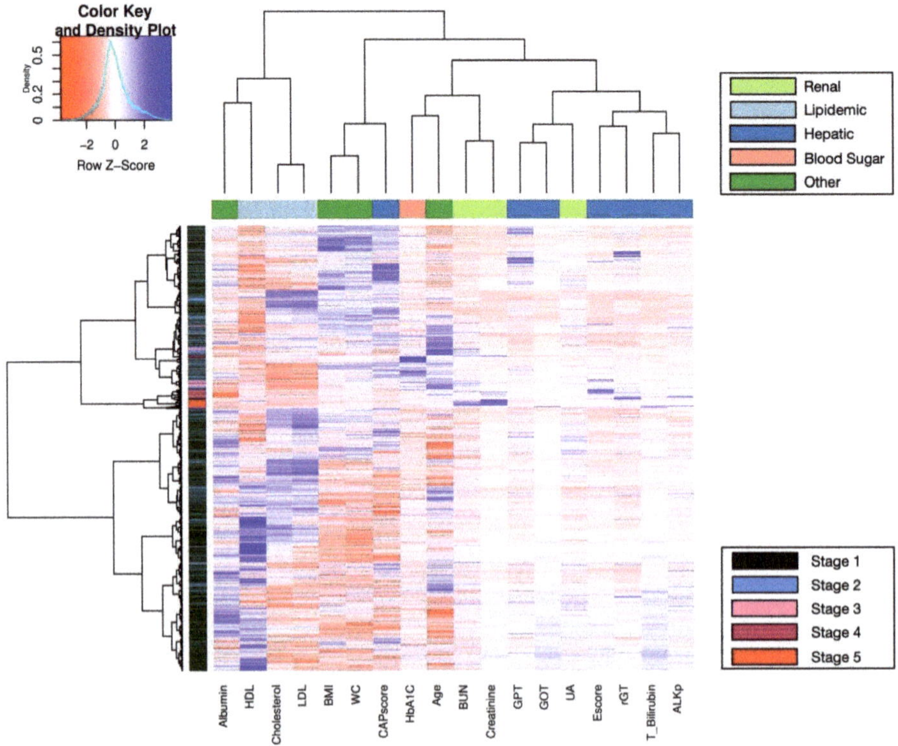

Figure 3. A clustering heatmap illustrating the classification of chronic kidney disease (CKD) in health examination data. Both the rows of CKD patients and the columns of biomarkers have been clustered, respectively; row data is also normalized into Z-score, simultaneously. Moreover, the mapping grids in the center have been colored according to their z scores.

4. Discussion

Traditionally, CKD stages have been defined by only eGFR, an approach that is neither punctilious nor precise, because CKD is a progressive and complex disease. This is especially complicated in early stage CKD patients with a steep decline in their renal function. Therefore, it is challenging for clinicians to predict the progression of CKD and to identify early patient groups at risk of rapid deterioration within the same stage of CKD. Using random forest analysis, we found that the significant variables that played principal roles in CKD were Creatinine, BUN and UA as nephritic factors; HbA1C as blood glucose factors; Escore, γGT and GPT as hepatic factors. In addition, this study found that a clustering heatmap was a practical and accurate method to classify CKD patients into several clusters as well as perform multigroup analysis. In addition, heatmaps that involved plenty of biomarkers were more dynamic, equitable, and comprehensive than a fixed criterion measured by MDRD considering only age, race, sex, and creatinine.

There is a paucity of studies that used machine learning to predict CKD, and we found one study that have very similar study design and results like ours. Using hospital data from India, Anusorn Charleonnan et al. found that the accuracy of prediction in CKD reached 0.98, very similar to what we have found (Figure 2b) [9]. However, Charleonnan et al. had not discussed and rank the significant factors, which contributed to prediction of CKD. Our study not only showed the importance of risk factors in predicting CKD by ensemble learning model, but also discussed the outcomes and meanings in clinical medicine. Furthermore, we have applied unsupervised learning technique, hierarchical clustering with heatmap visualization to classify and monitor patients in different groups. This approach might be useful for prediction and prevention of other chronic diseases that involved staging.

For heatmaps, we used Euclidean distance as the distance measure and Ward's method for unsupervised clustering. The clustering may be upgraded using other measures or algorithms. In addition, the heatmap may be more robust and dynamic with adequate sample size of patients at different stages or by using other specific and remarkable variables. Accordingly, new patients could be assessed using this clustering heatmap system based on their health conditions; if the new patients, especially those in early stages of CKD, had patterns similar to those in advanced CKD groups, precautionary treatment could be administered. In the clustering of medical factors, GPT, GOT and UA were in the same cluster with a similar pattern, whereas HDL, being a sign of good health, had the opposite pattern.

The CAP and E score by FibroScan are accurate non-invasive methods for assessing liver steatosis and fibrosis in patients with NAFLD [32]. CKD and NAFLD share a common pathological pathway and many important cardiometabolic risk factors. Moreover, the presence of pathophysiological interrelationships between the liver and kidney is well established and is supported by the presence of the hepatorenal syndrome, which may occur in patients with decompensated cirrhosis, regardless of its etiology. Patients with NAFLD had a significantly higher risk of incident CKD than those without NAFLD. Patients with more 'severe' NAFLD were also more likely to develop incident CKD [33]. NAFLD per se affects CKD through lipoprotein metabolism and hepatokine secretion, and conversely, targeting the renal tubule by sodium-glucose cotransporter 2 inhibitors can improve both CKD and NAFLD [34].

The result on hyperuricemia can be explained by association with the acceleration of GFR decline and CKD progression [35–37]. Previous research found that dyslipidemia in CKD is characterized by elevated triglycerides and decreased and dysfunctional HDL [38], and our results suggest that hypertension, hyperglycemia, and their combination may be associated with the incidence of CKD [39–41]. The association with HbA1C can be explained by diabetic patients being strongly associated with both albuminuria and reduced eGFR, and in the US it was found that diabetic patients had a substantially higher prevalence of CKD [42].

There are several limitations in this study. First, this study investigated only health-conscious participants that underwent a self-paid health examination, and the number of patients with advanced CKD may be underrepresented. Having approximately the same percentage of patients across different CKD stages will establish a much better model, and hence repeating this study by selectively including more advanced CKD patients is one of the future goals of this study. Second, this is a retrospective study, and confounding cannot be excluded. Hence, the subsequent step is to conduct a large prospective study to test the accuracy and usefulness of the predictive model. Third, this study concerned mainly Han Chinese population residing in Taiwan, and the study results needs to be repeated and validated in other populations. Fourth, this study failed to include some new CKD diagnostics that may further improve the prediction of CKD staging. For example, it was recently found that salivary ferric ion reducing antioxidant parameter can distinguish patients with mildly to moderately decreased kidney function from those with severe renal impairment with an accuracy of 100% (AUROC) [43].

5. Conclusions

Using machine learning technologies, we have generated a new predictive model of health care management for patients within the high-risk CKD group. However, considering the retrospective nature of this study, a sufficiently powered prospective cohort study is needed to conclusively address the usefulness of this predictive model to help physicians make an accurate diagnosis of the staging of CKD, which changes over time.

Author Contributions: Conceptualization, S.-S.C., C.-S.Y. and C.-H.L.; methodology, C.-S.Y.; software, C.-S.Y.; validation, C.-S.Y.; formal analysis, C.-S.Y.; investigation, C.-S.Y. and C.-H.L.; resources, C.-S.Y., C.-H.L., Y.-J.L., S.-Y.L., S.-T.W., J.L.W., M.-H.T., and S.-S.C.; data curation, C.-S.Y.; writing—original draft preparation, S.-S.C., C.-S.Y. and C.-H.L.; writing—review and editing, S.-S.C., C.-S.Y., C.-H.L. and J.L.W.; visualization, C.-S.Y.; supervision, S.-S.C.; project administration, S.-S.C., C.-S.Y. and C.-H.L.; funding acquisition, S.-S.C. All authors have read and agreed to the published version of the manuscript.

Funding: This research was funded by Taiwan National Science Foundation, grant number NSC108-2314-B-038-073.

Acknowledgments: We are grateful for the support from Taiwan National Science Foundation.

Conflicts of Interest: The authors declare no conflict of interest. The funders had no role in the design of the study; in the collection, analyses, or interpretation of data; in the writing of the manuscript, or in the decision to publish the results.

References

1. Nugent, R.A.; Fathima, S.F.; Feigl, A.B.; Chyung, D. The burden of chronic kidney disease on developing nations: A 21st century challenge in global health. *Nephron Clin. Pract.* **2011**, *118*, C269–C276. [CrossRef] [PubMed]
2. Hill, N.R.; Fatoba, S.T.; Oke, J.L.; Hirst, J.A.; O'Callaghan, C.A.; Lasserson, D.S.; Hobbs, F.D.R. Global prevalence of chronic kidney disease—A systematic review and meta-analysis. *PLoS ONE* **2016**, *11*, e158765. [CrossRef] [PubMed]
3. Wen, C.P.; Cheng, T.Y.D.; Tsai, M.K.; Chang, Y.C.; Chan, H.T.; Tsai, S.P.; Chiang, P.H.; Hsu, C.C.; Sung, P.K.; Hsu, Y.H.; et al. All-cause mortality attributable to chronic kidney disease: A prospective cohort study based on 462 293 adults in taiwan. *Lancet* **2008**, *371*, 2173–2182. [CrossRef]
4. Al-Aly, Z.; Zeringue, A.; Fu, J.; Rauchman, M.I.; McDonald, J.R.; El-Achkar, T.M.; Balasubramanian, S.; Nurutdinova, D.; Xian, H.; Stroupe, K.; et al. Rate of kidney function decline associates with mortality. *J. Am. Soc. Nephrol.: JASN* **2010**, *21*, 1961–1969. [CrossRef]
5. Coresh, J.; Turin, T.C.; Matsushita, K.; Sang, Y.; Ballew, S.H.; Appel, L.J.; Arima, H.; Chadban, S.J.; Cirillo, M.; Djurdjev, O.; et al. Decline in estimated glomerular filtration rate and subsequent risk of end-stage renal disease and mortality. *Jama* **2014**, *311*, 2518–2531. [CrossRef]
6. Matsushita, K.; Selvin, E.; Bash, L.D.; Franceschini, N.; Astor, B.C.; Coresh, J. Change in estimated gfr associates with coronary heart disease and mortality. *J. Am. Soc. Nephrol.: JASN* **2009**, *20*, 2617–2624. [CrossRef]
7. O'Hare, A.M.; Batten, A.; Burrows, N.R.; Pavkov, M.E.; Taylor, L.; Gupta, I.; Todd-Stenberg, J.; Maynard, C.; Rodriguez, R.A.; Murtagh, F.E.M.; et al. Trajectories of kidney function decline in the 2 years before initiation of long-term dialysis. *Am. J. Kidney Dis.* **2012**, *59*, 513–522.
8. Rosansky, S.J. Renal function trajectory is more important than chronic kidney disease stage for managing patients with chronic kidney disease. *Am. J. Nephrol.* **2012**, *36*, 1–10. [CrossRef]
9. Charleonnan, A.; Fufaung, T.; Niyomwong, T.; Chokchueypattanakit, W.; Suwannawach, S.; Ninchawee, N. Predictive analytics for chronic kidney disease using machine learning techniques. In Proceedings of the 2016 Management and Innovation Technology International Conference (MITicon), Bang-San, Chonburi, Thailand, 12–14 October 2016; pp. MIT-80–MIT-83.
10. Oeda, S.; Takahashi, H.; Imajo, K.; Seko, Y.; Ogawa, Y.; Moriguchi, M.; Yoneda, M.; Anzai, K.; Aishima, S.; Kage, M.; et al. Accuracy of liver stiffness measurement and controlled attenuation parameter using fibroscan® m/xl probes to diagnose liver fibrosis and steatosis in patients with nonalcoholic fatty liver disease: A multicenter prospective study. *J. Gastroenterol.* **2019**. [CrossRef]
11. Lin, Y.-J.; Lin, C.-H.; Wang, S.-T.; Lin, S.-Y.; Chang, S.-S. Noninvasive and convenient screening of metabolic syndrome using the controlled attenuation parameter technology: An evaluation based on self-paid health examination participants. *J. Clin. Med.* **2019**, *8*, 1775. [CrossRef]
12. Lee, J.I.; Lee, H.W.; Lee, K.S. Value of controlled attenuation parameter in fibrosis prediction in nonalcoholic steatohepatitis. *World J. Gastroenterol.* **2019**, *25*, 4959–4969. [CrossRef] [PubMed]
13. Ben Yakov, G.; Sharma, D.; Alao, H.; Surana, P.; Kapuria, D.; Etzion, O.; Rivera, E.; Huang, A.; Koh, C.; Heller, T.; et al. Vibration controlled transient elastography (fibroscan®) in sickle cell liver disease—Could we strike while the liver is hard? *Br. J. Haematol.* **2019**, *187*, 117–123. [CrossRef] [PubMed]

14. Vuppalanchi, R.; Chalasani, N. Nonalcoholic fatty liver disease and nonalcoholic steatohepatitis: Selected practical issues in their evaluation and management. *Hepatology* **2009**, *49*, 306–317. [CrossRef] [PubMed]

15. Alberti, K.; Zimmet, P.; Shaw, J. Metabolic syndrome—A new world-wide definition. A consensus statement from the international diabetes federation. *Diabetic Med.* **2006**, *23*, 469–480. [CrossRef]

16. Ma, Y.-C.; Zuo, L.; Chen, J.-H.; Luo, Q.; Yu, X.-Q.; Li, Y.; Xu, J.-S.; Huang, S.-M.; Wang, L.-N.; Huang, W.; et al. Modified glomerular filtration rate estimating equation for chinese patients with chronic kidney disease. *J. Am. Soc. Nephrol.: JASN* **2006**, *17*, 2937–2944. [CrossRef]

17. National Kidney Foundation. K/doqi clinical practice guidelines for chronic kidney disease: Evaluation, classification, and stratification. *Am. J. Kidney Dis. Off. J. Natl. Kidney Found.* **2002**, *39*, S1–S266.

18. Fox, J.; Monette, G. Generalized collinearity diagnostics. *J. Am. Stat. Assoc.* **1992**, *87*, 178–183. [CrossRef]

19. Fox, J.; Bates, D.; Firth, D.; Friendly, M.; Gorjanc, G.; Graves, S.; Heiberger, R.; Monette, G.; Nilsson, H.; Ripley, B.; et al. The car package. *R Found. Stat. Comput.* **2007**.

20. Fawcett, T. An introduction to roc analysis. *Pattern Recognit. Lett.* **2006**, *27*, 861–874.

21. Breiman, L.; Stone, C.J.; Friedman, J.H.; Olshen, R.A. *Classification and Regression Trees*; CRC Press: Boca Raton, FL, USA, 2017.

22. Therneau, T.M.; Atkinson, B.; Ripley, B. The RPART Package. 2010. Available online: https://cran.r-project.org/web/packages/rpart/index.html (accessed on 27 September 2019).

23. Breiman, L. Random forests. *Mach. Learn.* **2001**, *45*, 5–32. [CrossRef]

24. von Hippel, P.T. Biases in spss 12.0 missing value analysis. *Am. Stat.* **2004**, *58*, 160–164. [CrossRef]

25. Anderson, T.W. *An Introduction to Multivariate Statistical Analysis*; Wiley: New York, NY, USA, 1962.

26. Rokach, L.; Maimon, O. Clustering methods. In *Data mining and knowledge discovery handbook*; Springer US: Boston, MA, USA, 2005; pp. 321–352.

27. Ward, J.H. Hierarchical grouping to optimize an objective function. *J. Am. Stat. Assoc.* **1963**, *58*, 236–244. [CrossRef]

28. Cormack, R.M. A review of classification. *J. R. Stat. Soc. Ser. A* **1971**, *134*, 321–353. [CrossRef]

29. Wilkinson, L.; Friendly, M. The history of the cluster heat map. *Am. Stat.* **2009**, *63*, 179–184. [CrossRef]

30. Perrot, A.; Bourqui, R.; Hanusse, N.; Lalanne, F.; Auber, D. Large interactive visualization of density functions on big data infrastructure. In Proceedings of the 2015 IEEE 5th Symposium on large Data Analysis and Visualization (lDAV), Chicago, IL, USA, 25–26 October 2015; pp. 99–106.

31. Warnes, G.; Bolker, B.; Bonebakker, L.; Gentleman, R.; Huber, W.; Liaw, A.; Lumley, T.; Mächler, M.; Magnusson, A.; Möller, S. *Gplots: Various r Programming Tools for Plotting Data*. 2005. Available online: https://cran.r-project.org/web/packages/gplots/index.html (accessed on 27 September 2019).

32. Eddowes, P.J.; Sasso, M.; Allison, M.; Tsochatzis, E.; Anstee, Q.M.; Sheridan, D.; Guha, I.N.; Cobbold, J.F.; Deeks, J.J.; Paradis, V.; et al. Accuracy of fibroscan controlled attenuation parameter and liver stiffness measurement in assessing steatosis and fibrosis in patients with nonalcoholic fatty liver disease. *Gastroenterology* **2019**, *156*, 1717–1730. [CrossRef]

33. Mantovani, A.; Zaza, G.; Byrne, C.D.; Lonardo, A.; Zoppini, G.; Bonora, E.; Targher, G. Nonalcoholic fatty liver disease increases risk of incident chronic kidney disease: A systematic review and meta-analysis. *Metab. Clin. Exp.* **2018**, *79*, 64–76. [CrossRef]

34. Musso, G.; Cassader, M.; Cohney, S.; De Michieli, F.; Pinach, S.; Saba, F.; Gambino, R. Fatty liver and chronic kidney disease: Novel mechanistic insights and therapeutic opportunities. *Diabetes Care* **2016**, *39*, 1830–1845. [CrossRef]

35. Kuo, C.F.; Luo, S.F.; See, L.C.; Ko, Y.S.; Chen, Y.M.; Hwang, J.S.; Chou, I.J.; Chang, H.C.; Chen, H.W.; Yu, K.H. Hyperuricaemia and accelerated reduction in renal function. *Scand. J. Rheumatol.* **2011**, *40*, 116–121. [CrossRef]

36. Satirapoj, B.; Supasyndh, O.; Chaiprasert, A.; Ruangkanchanasetr, P.; Kanjanakul, I.; Phulsuksombuti, D.; Utainam, D.; Choovichian, P. Relationship between serum uric acid levels with chronic kidney disease in a southeast asian population. *Nephrology* **2010**, *15*, 253–258. [CrossRef]

37. Kang, D.-H.; Nakagawa, T. Uric acid and chronic renal disease: Possible implication of hyperuricemia on progression of renal disease. *Semin. Nephrol.* **2005**, *25*, 43–49. [CrossRef]

38. Reiss, A.B.; Voloshyna, I.; De Leon, J.; Miyawaki, N.; Mattana, J. Cholesterol metabolism in ckd. *Am. J. Kidney Dis.* **2015**, *66*, 1071–1082. [CrossRef] [PubMed]

39. Martinez-Castelao, A.; Navarro-Gonzalez, J.F.; Luis Gorriz, J.; de Alvaro, F. The concept and the epidemiology of diabetic nephropathy have changed in recent years. *J. Clin. Med.* **2015**, *4*, 1207–1216. [CrossRef] [PubMed]

40. Satirapoj, B.; Adler, S.G. Prevalence and management of diabetic nephropathy in western countries. *Kidney Dis.* **2015**, *1*, 61. [CrossRef] [PubMed]

41. Michishita, R.; Matsuda, T.; Kawakami, S.; Tanaka, S.; Kiyonaga, A.; Tanaka, H.; Morito, N.; Higaki, Y. Hypertension and hyperglycemia and the combination thereof enhances the incidence of chronic kidney disease (ckd) in middle-aged and older males. *Clin. Exp. Hypertens. (New York, N.Y.: 1993)* **2017**, *39*, 645–654. [CrossRef] [PubMed]

42. Zelnick, L.R.; Weiss, N.S.; Kestenbaum, B.R.; Robinson-Cohen, C.; Heagerty, P.J.; Tuttle, K.; Hall, Y.N.; Hirsch, I.B.; de Boer, I.H. Diabetes and ckd in the united states population, 2009-2014. *Clin. J. Am. Soc. Nephrol.: CJASN* **2017**, *12*, 1984–1990. [CrossRef]

43. Maciejczyk, M.; Szulimowska, J.; Taranta-Janusz, K.; Werbel, K.; Wasilewska, A.; Zalewska, A. Salivary frap as a marker of chronic kidney disease progression in children. *Antioxidants* **2019**, *8*, 409. [CrossRef]

Journal of
Clinical Medicine

Article

Robust Association between Acute Kidney Injury after Radical Nephrectomy and Long-Term Renal Function

Won Ho Kim [1,*], Kyung Won Shin [1], Sang-Hwan Ji [1], Young-Eun Jang [1], Ji-Hyun Lee [1], Chang Wook Jeong [2], Cheol Kwak [2] and Young-Jin Lim [1]

[1]   Department of Anesthesiology and Pain Medicine, Seoul National University Hospital, Seoul National University College of Medicine, Seoul 03080, Korea; skwskw1@naver.com (K.W.S.); taepoongshin@gmail.com (S.-H.J.); na0ag2@hotmail.com (Y.-E.J.); muslab@hanmail.net (J.-H.L.); limyjin@snu.ac.kr (Y.-J.L.)

[2]   Department of Urology, Seoul National University Hospital, Seoul National University College of Medicine, Seoul, 03080, Korea; drboss@korea.com (C.W.J.); mdrafael@snu.ac.kr (C.K.)

*   Correspondence: wonhokim@snu.ac.kr; Tel.: +82-2-2072-3484; Fax: +82-2-747-5639

Received: 9 January 2020; Accepted: 24 February 2020; Published: 25 February 2020

Abstract: The association between acute kidney injury (AKI) and long-term renal function after radical nephrectomy has not been evaluated fully. We reviewed 558 cases of radical nephrectomy. Postoperative AKI was defined by the Kidney Disease: Improving Global Outcomes (KDIGO) serum creatinine criteria. Values of estimated glomerular filtration rate (eGFR) were collected up to 36 months (median 35 months) after surgery. The primary outcome was new-onset chronic kidney disease (CKD) stage 3a or higher or all-cause mortality within three years after nephrectomy. The functional change ratio (FCR) of eGFR was defined as the ratio of the most recent GFR (24–36 months after surgery) to the new baseline during 3–12 months. A multivariable Cox proportional hazard regression analysis for new-onset CKD and a multivariable linear regression analysis for FCR were performed to evaluate the association between AKI and long-term renal outcomes. A correlation analysis was performed with the serum creatinine ratio and used to determine AKI and FCR. AKI occurred in 43.2% (n = 241/558) and our primary outcome developed in 40.5% (n = 226/558) of patients. The incidence of new-onset CKD was significantly higher in patients with AKI than those without at all follow-up time points after surgery. The Cox regression analysis showed a graded association between AKI and our primary outcome (AKI stage 1: Hazard ratio 1.71, 95% confidence interval 1.25–2.32; AKI stage 2 or 3: Hazard ratio 2.72, 95% confidence interval 1.78–4.10). The linear regression analysis for FCR showed that AKI was significantly associated with FCR (β = −0.168 ± 0.322, p = 0.011). There was a significant negative correlation between the serum creatinine ratio and FCR. In conclusion, our analysis demonstrated a robust and graded association between AKI after radical nephrectomy and long-term renal functional deterioration.

Keywords: acute kidney injury; radical nephrectomy; chronic kidney disease; functional change ratio

---

1. Introduction

Acute kidney injury (AKI) is a frequent complication after radical nephrectomy, with an incidence of up to 53.9% [1–3], and is associated with the development of chronic kidney disease (CKD) [4]. Although there are a few studies regarding this association after radical nephrectomy, the associations of AKI with new-onset CKD or long-term renal function after radical nephrectomy have not been evaluated fully [1–3]. To suggest a causal relationship, a dose-response relationship between AKI stages and CKD should be demonstrated [5]. However, the graded association of AKI stages with long-term renal function or the linear association of the ratio of creatinine used to diagnose AKI with a functional change ratio has not been evaluated before.

Several studies evaluated the association between AKI and new-onset CKD after radical nephrectomy. A previous retrospective study showed a significant association between AKI and new-onset CKD [2]. In this study, the new-onset CKD was defined as a 40% or more decrease in estimated glomerular filtration rate (eGFR) from the preoperative baseline. However, this study reviewed only a small number of 106 patients over a long period of eighteen years. Furthermore, there were 26% of patients with a baseline eGFR < 60 mL/min/1.73 m^2. These patients should have been excluded from their analysis. Another retrospective study reported AKI after radical nephrectomy is a risk factor for new-onset CKD [1]. This study defined CKD as eGFR < 60 mL/min/1.73 m^2, but 18% of patients with baseline eGFR < 60 mL/min/1.73 m^2 were included in the analysis. These two studies performed a logistic regression analysis to evaluate whether AKI was an independent risk factor of new-onset CKD. However, new-onset CKD is a time-to-event outcome and logistic regression analysis is not an adequate analysis for the outcome, hence both the Cox regression analysis and Kaplan–Meier survival analysis are required. In addition, mortality is a competing risk with new-onset CKD and mortality should be combined with the new-onset CKD for the Cox regression analysis.

As such, although a few previous studies have reported the association between AKI after nephrectomy and the risk of development of postoperative CKD, most studies analyzed a relatively small number of patients with different outcome definitions and used inadequate statistical analyses for the time-to-event outcome. As the appropriate statistical analyses were not selected, the linear relationship of the AKI stages with long-term renal functional change [6] or the graded association of AKI stages with CKD have not been evaluated. Therefore, we conducted a retrospective study evaluating the graded association of AKI stages with new-onset CKD and the linear association of AKI stages with renal functional change ratio from new-baseline renal function after radical nephrectomy to assess whether the association of AKI and CKD after radical nephrectomy is robust and has a dose-response relationship. We excluded the patient who received donor nephrectomy for kidney transplantation from our analysis as these patients have different renal functional profiles after surgery. This difference is attributed to the different distribution of patient age and comorbidities [7,8].

2. Methods

2.1. Patient Selection

This single-center retrospective observational study was approved by the institutional review board (IRB) of Seoul National University Hospital (1907-172-1050). Written informed consent was waived by the IRB due to the retrospective nature of the present study. We reviewed electronic medical records of the patients who were ≥ 20 years old, had a renal mass, and underwent radical nephrectomy, regardless of surgical techniques between 2010 and 2015. Among the 670 patients who underwent radical nephrectomy, the patients with baseline eGFR < 60 mL/min/1.73 m^2 (n = 50), the absence of contralateral kidney (n = 9), or missing baseline serum creatinine (n = 0) or three or more follow-up loss of serum creatinine or eGFR values among 3, 12, 24, and 36 months after surgery (n = 22) were excluded. We also further excluded patients who underwent subsequent nephrectomy (n = 2), catheter ablation of remaining kidney (n = 5), chemotherapy with a nephrotoxic agent (n = 4), or received other nephrotoxic drugs including diuretics (n = 3), nephrotoxic antibiotics (n = 4), and radiocontrast media (n = 13) during our follow-up period because these could affect residual renal function. The remaining 558 patients were included in the final analysis.

2.2. Patient Data and Outcome Measurements

Demographic, baseline characteristics, and surgery-related data which were known to be associated with renal function after nephrectomy were obtained from our electronic medical records (Table 1) [1–3,9]. Serum creatinine values measured at 3, 12, 24, 36 months after surgery were collected. GFR values were estimated at these time points after surgery by the Modification in Diet and Renal Disease (MDRD) study equation [10]. Decisions regarding radical nephrectomy and the type of surgical approach were made by surgeons based on tumor characteristics.

Table 1. Patient characteristics and perioperative parameters according to acute kidney injury (*n* = 558).

Variables	Patients	Proportion without Missing (%)
Demographic data		
Age, year	60 (51–68)	100
Female, *n*	171 (30.6)	100
Body-mass index, kg/m^2	24.4 (22.6–26.2)	100
Baseline medical status		
Hypertension, *n*	287 (51.4)	100
Diabetes mellitus, *n*	98 (17.6)	100
Cerebrovascular accident, *n*	11 (2.0)	100
Angina pectoris, *n*	9 (1.6)	100
Preoperative hemoglobin, g/dL	13.5 (12.0–14.6)	100
Preoperative serum albumin level, mg/dL	4.3 (4.1–4.6)	99.8
Preoperative proteinuria, *n*	67 (12.0)	100
Preoperative hematuria, *n*	53 (9.5)	100
Preoperative eGFR, mL/min/1.73 m^2		100
eGFR ≥ 90 mL/min/m^2	153 (27.4)	
eGFR 60–89 mL/min/1.73m^2	405 (72.6)	
Surgical parameters		
Surgery type, *n*		100
Laparoscopic	223 (40.0)	
Robot-assisted	9 (1.6)	
Open	325 (58.2)	
Clinical stage, *n*		100
T1a/ T1b	141 (25.3)/152 (27.2)	
T2a/ T2b	154 (27.6)/61 (10.9)	
T3a/T3b/T3c	19 (3.4)/ 17 (3.0)/14 (2.5)	
N 0/1	514 (92.1)/44 (7.9)	
M 0/1	520 (93.2)/38 (6.8)	
R.E.N.A.L. score		100
Low (4–6)	225 (40.3)	
Intermediate (7–9)	286 (51.3)	
High (10–12)	47 (8.4)	
Tumor maximal diameter, cm	5.5 (3.2–7.8)	100
Operation time, min	130 (100–170)	100
Bleeding and transfusion amount		
pRBC transfusion, *n*	52 (9.3)	100
Estimated blood loss, mL	200 (100–400)	99.6
Anesthesia-related parameters		
Volatile anesthetics use, *n*	494 (88.5)	
Total intravenous anesthesia, *n*	64 (11.5)	
Crystalloid administration, mL	1100 (750–1500)	100
Colloid administration, mL	0 (0–300)	100
Vasopressor infusion during surgery	29 (5.2)	100

Data are presented as median (interquartile range) or number (%). AKI = acute kidney injury; eGFR = estimated glomerular filtration rate; R.E.N.A.L. = radius, exophytic/endophytic properties, nearness of tumor to collecting system or sinus, anterior/posterior, hilar, location relative to polar lines; pRBC = packed red blood cell.

The primary outcome was new-onset CKD or higher or all-cause mortality within three years after radical nephrectomy. The new-onset CKD was defined as a decrease in eGFR < 60 mL/min/1.73 m^2 or the initiation of chronic hemodialysis [11]. The decrease in eGFR should be identified by at least two consecutive measurements separated by an interval of at least three months [12]. The primary outcome variable was defined as a time-to-event outcome.

Secondary outcomes included the long-term functional change ratio (FCR) of GFR, which was defined as the most recent GFR/new baseline GFR after surgery [6]. New baseline GFR was defined as the latest value available during 3–12 months after surgery considering that renal function recovers after the initial drop immediately after surgery. The most recent GFR value of at least 24 months after surgery was compared to this new baseline.

Postoperative AKI was defined by the creatinine criteria of Kidney Disease: Improving Global Outcomes (KDIGO), which was determined according to the maximal change in serum creatinine levels during the first seven postoperative days (stage 1: 1.5–1.9; stage 2: 2–2.9; stage 3: More than 3-fold increase of baseline, respectively, also stage 1: when serum creatinine increased by 0.3 mg/dL within 48 h) [13,14]. The most recent serum creatinine level measured before surgery was used as the baseline.

2.3. Statistical Methods

SPSS software version 25.0 (IBM Corp., Armonk, NY, USA) and STATA/MP version 15.1 (StataCorp, College Station, TX, USA) were used to analyze the data. The visual inspection of histograms and quintile–quintile plots was performed to determine the normality of the data. Continuous variables were presented as mean ± SD for normally distributed data or median (25th, 75th percentiles) for non-normally distributed data. For all analyses, $p < 0.05$ was considered statistically significant. Baseline characteristics and surgery-related parameters had missing values in <1% and these missing were considered to be due to missing completely at random. The incidences of missing in the baseline parameters were reported and they were not replaced. Missing values of serum creatinine and eGFR at 3, 12, 24, 36 months were reported and were not replaced, and an analysis was performed with available data. One of our investigators (W.K.) verified artifact or incorrect values according to our source document and excluded them from our analysis. The following are a summary of our main analyses. Firstly, the serial eGFR values up to 36 months after surgery and cumulative incidences of postoperative development of new-onset CKD were depicted and compared at all follow-up time points between the patients without AKI, those with AKI stage 1, and those with AKI stage 2 or 3. The absolute eGFR values were compared among no-AKI, AKI stage 1, and stage 2 or 3 at each time point by one-way analysis of variance. The cumulative incidence of new-onset CKD was compared between the same groups by the chi-square test. Bonferroni correction was used to adjust for the increased type 1 error by multiple comparisons ($p < 0.05/4 = 0.012$).

Secondly, we scrutinized the relationship between postoperative AKI and the development of new-onset CKD during the 60 months after surgery through a multivariable Cox proportional hazard regression analysis. Proportional hazard assumptions were tested by visual inspection of log-minus-log survival plots for categorical variables and restricted cubic splines for continuous variables [15,16]. Before conducting multivariable analysis, multicollinearity among covariates was evaluated using the variance inflation factor. Variables with variance inflation factor > 5 were excluded from the analysis. Cases with missing values of the covariates were excluded from the Cox regression analysis for the complete case analysis. All collected perioperative variables listed in Table 1 were entered into the multivariable model to adjust for the association between AKI and CKD. AKI as a binary variable or the stages of AKI an ordinal variable were entered alternatively as a covariate in the regression model. All covariates entered the model without stepwise variable selection or a univariable screening. Calibration of the Cox regression model was evaluated by Gronnesby and Borgan test [17] and discrimination of the model was measured by Harrell's *c* and Somers' *D* [18].

Thirdly, a Kaplan–Meier survival curve analysis of our primary outcome was performed among the different groups of the patients without AKI, those with AKI stage 1, and those with AKI stage 2 or 3. The log-rank test was used to determine statistical significance between groups.

Fourthly, as analysis for the secondary outcome of FCR, we performed a multivariable linear regression analysis to elucidate whether AKI is significantly associated with the long-term FCR of the most recent eGFR to a new postoperative renal functional baseline. Linear regression model assumptions were evaluated using residual plots and scatter plots of our data. Neither stepwise variable selection nor univariable screening was performed. Multicollinearity among covariates was evaluated using the variance inflation factor.

Fifthly, a scatter plot was depicted relating the distribution of FCR across the serum creatinine ratio used to determine AKI. We tested whether the distribution of FCR is different among the different stages of AKI by univariable Spearman correlation analysis.

Sixthly, as a post-hoc analysis to evaluate the postoperative renal function in terms of KDIGO CKD stages [12], the distribution of KDIGO CKD stages during the follow-up period was compared between those with and without AKI.

Finally, as a post-hoc analysis to evaluate the association of the type of surgical procedure and AKI, we compared the incidence of AKI between open surgery and laparoscopic or robotic surgery. In addition, to evaluate the association between the surgical procedure and our primary outcome, we performed Kaplan–Meier survival curve analysis. The log-rank test was used to determine statistical significance.

3. Results

Demographics and perioperative parameters were shown in Table 1. The incidence of AKI was 43.2% ($n = 241/558$) (stage 1: $n = 160$ (28.7%); stage 2 or 3: $n = 81$ (14.5%)). Our primary outcome developed in 40.5% ($n = 226/558$) of patients. The median follow-up of renal function was 35 (25–37) months. The numbers of patients lost to follow-up at 2 and 3 years after surgery were 143 (24.7%) and 305 (52.8%) (Table S1).

Table 1 shows the baseline characteristics of the patients. Median patient age was 60 years and 60.7% of patients were male. Median FCR at three years after radical nephrectomy was 0.94 in patients without AKI and 0.80/0.72 for patients with stage 1/2 or 3 AKI.

Figure 1 shows the comparison of a time-dependent change in eGFR at 3 months, 1, 2, and 3 years after radical nephrectomy among the patients without AKI, those with stage 1 AKI and those with stage 2 or 3 AKI. There were significant differences in eGFR values between groups in all time points ($p < 0.001$). Figure 2 shows comparisons of the cumulative incidence of new-onset CKD at the time points of follow-up between those with and without AKI. While 24.7% of patients with no history of AKI developed CKD until three years after surgery, 69.5% of patients developed CKD with a history of stage 2 or 3 AKI ($p < 0.001$). There were significant differences in the incidence of new-onset CKD at all time points ($p < 0.001$).

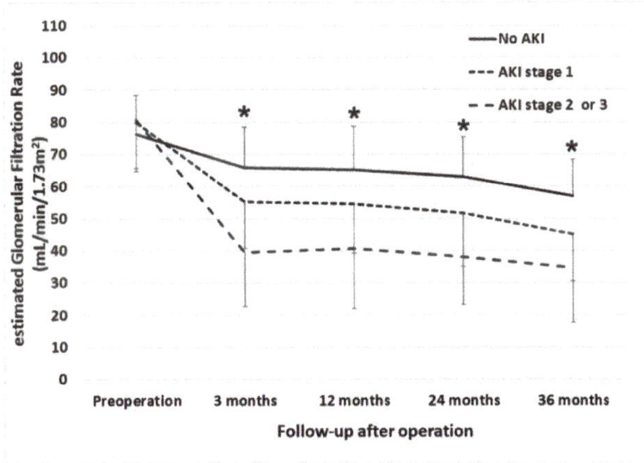

Figure 1. Serial change in mean estimated glomerular filtration rate before surgery to three years after surgery according to the stages of acute kidney injury. * Significant difference between groups.

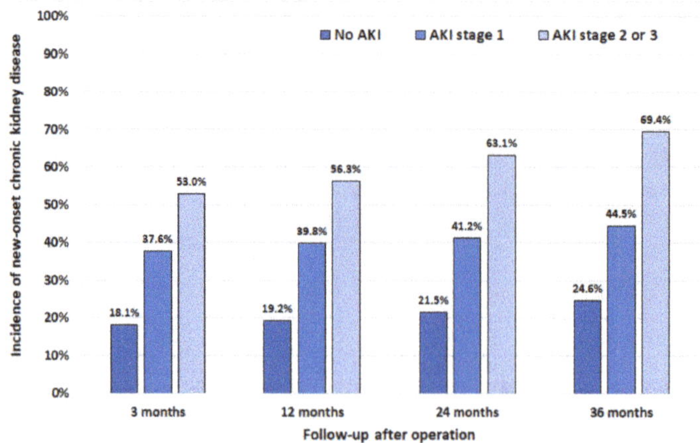

Figure 2. Comparison of the cumulative incidence of chronic kidney disease (estimated glomerular filtration rate (eGFR) <60 mL/min/1.73 m^2) at follow-up time points between the patients without acute kidney injury (AKI), those with stage 1 AKI, and those with stage 2 or 3 AKI. There were significant differences in the incidence of chronic kidney disease at all time-points.

Table 2 shows the results of the multivariable Cox regression analysis to predict the development of new-onset CKD during the three years after radical nephrectomy. All covariates met the proportional hazard assumptions. Postoperative AKI was identified as an independent risk factor for the development of new-onset CKD (multivariable-adjusted hazard ratio (HR) 2.46, 95% confidence interval (CI) 1.70–3.63, $p < 0.001$). There was a graded association with CKD for stage 1 (HR 1.71, 95% CI 1.25-2.32, $p < 0.001$) and stage 2 or 3 AKI (HR 2.72, 95% CI 1.78–4.10, $p < 0.001$). The performance of our multivariable Cox model in terms of Harrell's c and Somers' D was 0.67 and 0.60, respectively. Our Cox regression model showed good calibration (Gronnesby and Borgan test: $\chi^2 = 0.876$, $p = 0.441$).

Table 2. Cox proportional hazard regression analysis for new-onset chronic kidney disease or all-cause mortality during the three years after radical nephrectomy in all patients ($n = 558$).

Variable	Hazard Ratio	95% CI	p-Value
Age, years	1.05	1.00–1.09	0.043
Female	1.30	0.81–2.10	0.368
Body-mass index, kg/m^2	1.01	0.95–1.08	0.769
History of hypertension	1.70	1.07–2.78	0.022
History of diabetes mellitus	1.95	1.13–3.44	0.012
Preoperative hemoglobin, g/dL	1.14	0.99–1.30	0.064
Preoperative albumin, g/dL	0.63	0.33–1.12	0.077
Preoperative proteinuria, n	0.82	0.42–1.80	0.547
Preoperative hematuria, n	1.12	0.57–1.74	0.657
Preoperative estimated glomerular filtration rate, mL/min/1.73m^2	0.99	0.98–0.99	0.042
Postoperative acute kidney injury	2.46	1.70–3.63	<0.001
No acute kidney injury	baseline		
Acute kidney injury stage 1	1.71	1.25–2.32	<0.001
Acute kidney injury stage 2 or 3	2.72	1.78–4.10	<0.001
Preoperative tumor maximal diameter, cm	1.05	0.97–1.12	0.164
Open surgery (vs. laparoscopic or robot-assisted)	0.74	0.49–1.15	0.255
Operation time, hour	0.96	0.81–1.18	0.847

Table 2. Cont.

Variable	Hazard Ratio	95% CI	*p*-Value
Total intravenous anesthesia	0.89	0.61–1.35	0.558
Intraoperative crystalloid administration, per 100 mL	0.87	0.62–1.28	0.415
Intraoperative colloid administration, per 100 mL	1.06	0.98–1.16	0.176
Intraoperative vasopressor infusion, *n*	0.94	0.92–1.17	0.514
Red blood cell transfusion, *n*	0.82	0.37–1.75	0.427

CI = confidence interval. Intraoperative vasopressor infusion means norepinephrine or phenylephrine infusion during surgery.

Figure 3 shows Kaplan–Meier survival curve analysis for our primary outcome between different AKI groups. There were significant differences between the no AKI and AKI stage 1 groups (log-rank test $p < 0.001$) and between AKI stage 1 and stage 2 or 3 (log-rank test $p = 0.023$). The numbers of patients who had follow-up eGFR values at each time point were shown in Figure 3.

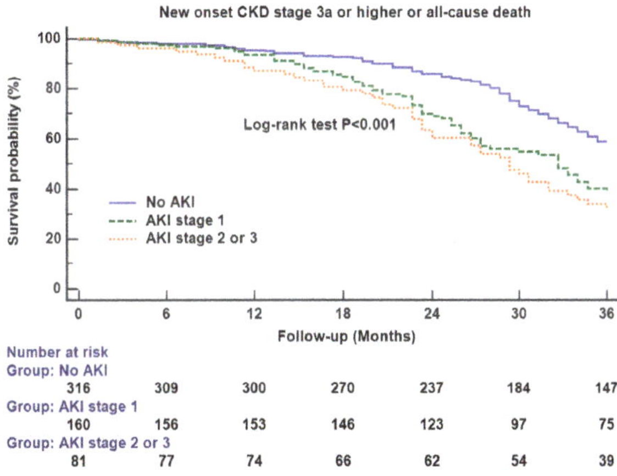

Figure 3. Kaplan–Meier survival curve analysis for new-onset chronic kidney disease stage 3a or higher (estimated glomerular filtration rate < 60 mL/min/1.73 m^2) or all-cause mortality as the primary outcome between different AKI groups. There were significant differences between the no AKI and AKI stage 1 groups (log-rank test $p < 0.001$) and between AKI stage 1 and stage 2 or 3 (log-rank test $p = 0.023$). The numbers of patients who had follow-up eGFR values at each time point were shown at the bottom of the figure.

Table 3 shows the results of the multivariable linear regression analysis for FCR after radical nephrectomy. All covariates met the linear assumptions. Postoperative AKI was independently associated with the FCR of the most recent follow-up ($\beta = -0.168 \pm 0.322$, $p = 0.011$). The performance of our multivariable prediction in terms of R^2 was 0.28 and there was no significant multicollinearity between covariates.

Figure 4 shows the distribution of FCR across the serum creatinine ratio used to determine AKI and AKI stages. There was a significant negative correlation between the FCR and the serum creatinine ratio ($p < 0.001$). This means the residual renal function deteriorates more as the stages of AKI increase.

Figure 5 compares the distribution of the KDIGO CKD stages during our follow-up period of 36 months between those with and without AKI. There was a significant difference in the distribution of CKD stages between the patients with and without AKI.

Table 3. Multivariable linear regression analysis of functional change ratio after radical nephrectomy.

Variable	$\beta \pm$ Standard Error	*p*-Value	VIF
Age, years	0.012 ± 0.001	0.057	1.69
Female	0.037 ± 0.031	0.240	1.30
Body-mass index, kg/m^2	0.002 ± 0.005	0.647	1.14
History of hypertension	-0.030 ± 0.011	0.047	1.52
History of diabetes mellitus	-0.044 ± 0.018	0.044	1.24
Preoperative hemoglobin concentration, g/dL	0.007 ± 0.010	0.411	2.06
Preoperative albumin level, mg/dL	0.033 ± 0.041	0.481	1.82
Preoperative proteinuria	-0.034 ± 0.047	0.470	1.45
Preoperative estimated glomerular filtration rate, per 10 mL/min/1.73 m^2	0.170 ± 0.122	0.002	1.26
Postoperative acute kidney injury	-0.168 ± 0.322	0.011	1.16
Maximal diameter of renal mass, cm	0.002 ± 0.004	0.572	1.28
Open surgery (vs. laparoscopic or robot-assisted)	-0.030 ± 0.028	0.228	1.19
Operation time, hour	-0.016 ± 0.012	0.179	1.20
Total intravenous anesthesia	0.069 ± 0.056	0.375	1.10
Intraoperative crystalloid administration, mL/kg	-0.011 ± 0.023	0.724	1.45
Intraoperative colloid administration, mL/kg	-0.001 ± 0.004	0.717	1.37
Intraoperative red cell transfusion, n	0.096 ± 0.049	0.150	1.39

VIF = variance inflation factor. The functional change ratio was determined as a ratio of the most recent estimated glomerular filtration rate (eGFR) (at least 24 months and up to three years after surgery) to eGFR at 3 to 12 months.

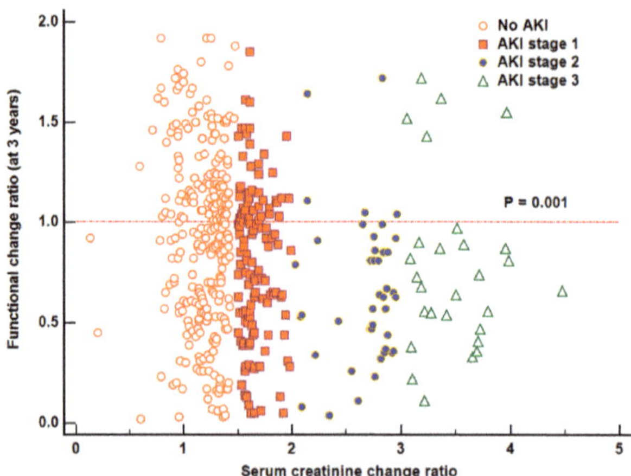

Figure 4. Distribution of the long-term functional change ratio across the different creatinine ratios which were used to determine acute kidney injury stages. There was a significant correlation between the functional change ratio and creatinine change ratio from baseline ($p < 0.001$).

The incidence of postoperative AKI was significantly lower in the laparoscopic or robotic surgery group ($n = 77$, 32.6%) compared to the open surgery group ($n = 169$, 49.4%, $p < 0.001$) (Table S2). However, there was no significant difference in the renal survival between the laparoscopic or robotic group and the open group (Log-rank test $p = 0.865$) (Figure S1). As a supplementary analysis, the incidence and type of postoperative complication were compared between the patients with and without AKI (Table S3). There was no significant difference between groups.

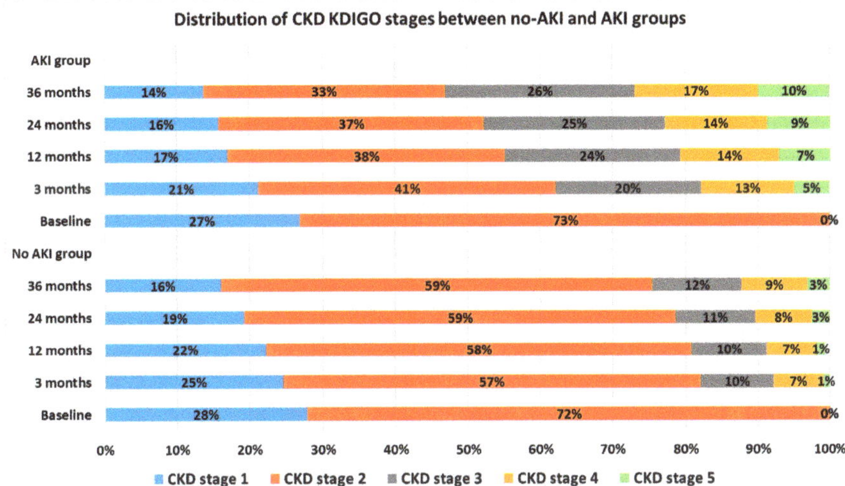

Figure 5. Comparison of the distribution of the Kidney Disease: Improving Global Outcomes chronic kidney disease (KDIGO CKD) stages during the follow-up period of 36 months between the patients with and without AKI after radical nephrectomy.

4. Discussion

We evaluated the association between AKI after radical nephrectomy and long-term renal function through a multivariable Cox regression analysis and other sensitivity analyses in a retrospective cohort during a 36-months follow-up. The association between AKI and CKD was revealed again in this specific surgical setting of radical nephrectomy. Furthermore, we demonstrated the graded associations between AKI stages and new-onset CKD as well as between AKI stages and the FCR at 24–36 months after surgery. Our linear regression analysis also showed that AKI was an independent predictor of FCR. Therefore, we added a convincing body of evidence that there is a robust and linear relationship between AKI and long-term renal function after radical nephrectomy. Given the high incidence of new-onset CKD after radical nephrectomy in patients with AKI, clinical trials are urgently required to find a modifiable risk factor to mitigate the risk of AKI.

According to the previous studies comparing radical and partial nephrectomy, radical nephrectomy is more strongly associated with new-onset CKD after surgery [19–22]. A recent meta-analysis compared the postoperative incidence of CKD stage 3a or higher (eGFR < 60 mL/min/1.73 m^2) between radical and partial nephrectomy and reported a higher risk of CKD for radical nephrectomy [23]. However, these studies simply compared renal functional outcomes between radical and partial nephrectomy and cannot confirm the causal relationship between radical nephrectomy and new-onset CKD. A recent study constructed a simplified nomogram that predicts postoperative renal functional decline after nephrectomy and could help determine candidates for partial nephrectomy [24]. For stage 1 renal tumors, oncologic outcomes were not different between partial and radical nephrectomy [25,26]. However, partial nephrectomy is associated with a lower incidence of postoperative renal functional decline and lower mortality compared to radical nephrectomy [27,28]. Therefore, nephron-sparing surgery has been proposed as a standard of care for small renal cell carcinomas [29]. However, radical nephrectomy is still an important option when partial nephrectomy is unfeasible [30].

In surgeries other than nephrectomy, AKI has been reported to be closely associated with the development of CKD and increased mortality [4,31]. The association between AKI and CKD was also reported in non-surgical patients [32,33]. However, this association was not found in patients with a history of septic shock during one-year follow-up [34]. Furthermore, there have been only a few studies investigating the association between AKI and long-term renal functional decline for radical nephrectomy. Based on our literature review, there have been three studies that investigated the association between postoperative AKI and the development of progressive chronic kidney disease [1–3]. The incidence of AKI was reported to be 34% to 49% and the association between AKI and CKD was strong, with a three to four-fold higher risk pertaining to the odds ratio, although the CKD was defined differently with varying durations of follow-up. The incidence of AKI in our study was higher than in the study of Cho et al., in which it was 33.7% [1], but lower than that of Garfalo et al. at 49.1% [2]. The highest incidence of 53.9% was reported in patients undergoing radical nephrectomy with simultaneous inferior vena cava (IVC) thrombectomy [3]. This high incidence may be due to the frequent implementation of cardiopulmonary bypass and clamping of IVC and the contralateral renal vein. The variance in the incidence of AKI may also be attributed to different diagnostic criteria and the exclusion of urine output criteria. We did not use the urine output criteria because hourly urine output was not accurately recorded in our patients and the oliguria cutoff associated with AKI could be different in the perioperative setting [35,36]. Urine output criteria are regarded as unreliable in predicting AKI in the perioperative environment because oliguria may develop simply due to decreased preload [37], or external obstruction of the urinary tract [38].

Renal function after donor nephrectomy was ascertained in previous studies, which could be compared with our study results [39–41]. Compensatory hypertrophy develops in the remaining kidney [39–41]. Serum creatinine levels increased to 20% above baseline but usually remained within normal reference range [7,42]. However, in our study, immediate postoperative AKI developed in 43.2% and more than half of our patients with AKI eventually developed CKD of eGFR <60 mL/min/1.73 m^2 during three years of follow-up. These differences in renal outcomes between kidney donors and our patients with renal cell carcinoma could be attributed to patient characteristics. Our cancer patients are older and have more comorbidities [7,8], which could aggravate the renal functional decline. Our incidence of new-onset CKD during 35 months of median follow-up after radical nephrectomy is similar to previous studies of 38% [1] or 39.6% [2], although the follow-up duration differed. Another previous study reported a higher incidence of 65%, although this study included 26% of patients with pre-existing CKD before surgery [21].

Since AKI after radical nephrectomy is associated with long-term renal functional decline, efforts should be made to reduce the risk of AKI after radical nephrectomy. Creatinine increase after radical nephrectomy would be due to both losses of nephrons in the resected kidney and injury of the contralateral kidney. Although we cannot reduce the loss of nephrons in patients for whom nephron-sparing surgery is not feasible, injury of the contralateral kidney could be prevented by hemodynamic optimization during the perioperative period [43]. Previously reported risk factors of male sex, old age, high body-mass index, hypertension, diabetes mellitus, preoperative proteinuria, high preoperative eGFR are not modifiable [1,2,21,44–46]. Comorbidities such as hypertension or diabetes could contribute to the underlying medical renal disease and long-term renal functional deterioration [47]. Our analysis also identified unmodifiable predictors, such as preoperative hypertension, diabetes mellitus, and low baseline eGFR. A previous pilot study reported that remote ischemic conditioning reduced the incidence of AKI in patients undergoing partial nephrectomy [48]. Further studies are required to find modifiable risk factors of AKI or CKD progression after radical nephrectomy [9] and determinants for adaptive hyperfiltration in the contralateral kidney [39].

All surgical procedures including open, laparoscopic, and robotic surgery were pooled in our analysis. A previous study showed that postoperative risk of AKI was halved for the patients treated by robotic or laparoscopic partial nephrectomy when compared with open surgery [49]. To further evaluate this association in our study cohort, we compared the incidence of AKI and the renal survival between the groups of surgical procedure. The incidence of AKI was significantly lower in the laparoscopic or robotic surgery group compared to open surgery. However, there was no significant difference in our primary outcome according to the type of surgical procedure both for Cox regression analysis and survival curve analysis. This may suggest that a surgical procedure may affect short-term outcomes but not long-term renal survival after radical nephrectomy.

Our study should be interpreted under several limitations. Firstly, our study was a retrospective study of a single tertiary care center. Minimally invasive and open nephrectomy were mixed in our population. Although a multivariable analysis and sensitivity analyses were performed, unknown and unmeasured biases could have affected our results. Secondly, there was about 50% follow-up loss of renal function at 36 months after surgery, as shown in Figure 3. A prospective observational study with less loss of follow-up is required for a more valid analysis. Thirdly, our study cohort included patients who might have been considered to undergo partial nephrectomy simply according to the tumor stage provided in Table 1. This may limit the applicability of our data to the current clinical practice. However, other indications of radical nephrectomy such as locally advanced cases, multifocality, and distant metastasis should be considered for this. Fourthly, the exact causes of new-onset CKD during our follow-up period were not elucidated. Although the main cause would be functional declines in the remaining kidney, other pathologies such as new-onset glomerulonephritis could have been included in our primary outcome.

5. Conclusions

AKI after radical nephrectomy was associated with long-term renal functional deterioration in a dose-response manner. AKI was an independent risk factor of new-onset CKD in patients without baseline CKD. There was a significant linear relationship between the creatinine ratio used to determine AKI and FCR at 24–36 months after surgery. These associations were stronger in higher stages of AKI. Therefore, there seems to be a robust and strong association between AKI after radical nephrectomy and long-term renal functional decline after surgery. It is required to find modifiable risk factors of AKI and interventions to attenuate AKI to improve long-term renal function after radical nephrectomy.

Supplementary Materials: The following is available online http://www.mdpi.com/2077-0383/9/3/619/s1. Figure S1. Kaplan–Meier survival curve analysis for new-onset chronic kidney disease stage 3a or higher (estimated glomerular filtration rate < 60 mL/min/1.73 m^2) or all-cause mortality as the primary outcome between different type of surgical procedure groups; Table S1. The number of patients who had follow-up values of renal function; Table S2. Comparison of the incidence of acute kidney injury between the types of surgical procedures; Table S3. Comparison of the incidence of complications between the patients with and without acute kidney injury according to Clavien-Dindo classification.

Author Contributions: W.H.K.: conceptualization, data curation, formal analysis, investigation, methodology, visualization, and writing-original draft; K.W.S.: data curation, formal analysis, writing-review and editing; S.-H.J.: writing-review and editing; Y.-E.J.: writing-review and editing; J.-H.L.: writing-review and editing; C.W.J.: data curation, writing-review and editing; C.K.: writing-review and editing; Y.-J.L.: writing-review and editing. All authors have read and agreed to the published version of the manuscript.

Funding: No external fund received.

Conflicts of Interest: The authors declare no conflict of interest.

References

1. Cho, A.; Lee, J.E.; Kwon, G.Y.; Huh, W.; Lee, H.M.; Kim, Y.G.; Kim, D.J.; Oh, H.Y.; Choi, H.Y. Post-operative acute kidney injury in patients with renal cell carcinoma is a potent risk factor for new-onset chronic kidney disease after radical nephrectomy. *Nephrol. Dial. Transpl.* **2011**, *26*, 3496–3501. [CrossRef]
2. Garofalo, C.; Liberti, M.E.; Russo, D.; Russo, L.; Fuiano, G.; Cianfrone, P.; Conte, G.; De Nicola, L.; Minutolo, R.; Borrelli, S. Effect of post-nephrectomy acute kidney injury on renal outcome: A retrospective long-term study. *World J. Urol.* **2018**, *36*, 59–63. [CrossRef]
3. Shin, S.; Han, Y.; Park, H.; Chung, Y.S.; Ahn, H.; Kim, C.S.; Cho, Y.P.; Kwon, T.W. Risk factors for acute kidney injury after radical nephrectomy and inferior vena cava thrombectomy for renal cell carcinoma. *J. Vasc. Surg.* **2013**, *58*, 1021–1027. [CrossRef] [PubMed]
4. Chawla, L.S.; Eggers, P.W.; Star, R.A.; Kimmel, P.L. Acute kidney injury and chronic kidney disease as interconnected syndromes. *N. Engl. J. Med.* **2014**, *371*, 58–66. [CrossRef] [PubMed]

5. Hill, A.B. The Environment and disease: Association or causation? *Proc. R. Soc. Med.* **1965**, *58*, 295–300. [CrossRef] [PubMed]

6. Zabell, J.; Isharwal, S.; Dong, W.; Abraham, J.; Wu, J.; Suk-Ouichai, C.; Palacios, D.A.; Remer, E.; Li, J.; Campbell, S.C. Acute Kidney Injury after Partial Nephrectomy of Solitary Kidneys: Impact on Long-Term Stability of Renal Function. *J. Urol.* **2018**, *200*, 1295–1301. [CrossRef]

7. Miller, I.J.; Suthanthiran, M.; Riggio, R.R.; Williams, J.J.; Riehle, R.A.; Vaughan, E.D.; Stubenbord, W.T.; Mouradian, J.; Cheigh, J.S.; Stenzel, K.H. Impact of renal donation. Long-term clinical and biochemical follow-up of living donors in a single center. *Am. J. Med.* **1985**, *79*, 201–208. [CrossRef]

8. Ibrahim, H.N.; Foley, R.; Tan, L.; Rogers, T.; Bailey, R.F.; Guo, H.; Gross, C.R.; Matas, A.J. Long-term consequences of kidney donation. *N. Engl. J. Med.* **2009**, *360*, 459–469. [CrossRef]

9. Lee, H.J.; Bae, J.; Kwon, Y.; Jang, H.S.; Yoo, S.; Jeong, C.W.; Kim, J.T.; Kim, W.H. General Anesthetic Agents and Renal Function after Nephrectomy. *J. Clin. Med.* **2019**, *8*, 1530. [CrossRef]

10. Levey, A.S.; Bosch, J.P.; Lewis, J.B.; Greene, T.; Rogers, N.; Roth, D. A more accurate method to estimate glomerular filtration rate from serum creatinine: A new prediction equation. Modification of Diet in Renal Disease Study Group. *Ann. Intern. Med.* **1999**, *130*, 461–470. [CrossRef]

11. Levey, A.S.; Coresh, J.; Balk, E.; Kausz, A.T.; Levin, A.; Steffes, M.W.; Hogg, R.J.; Perrone, R.D.; Lau, J.; Eknoyan, G. National Kidney Foundation practice guidelines for chronic kidney disease: Evaluation, classification, and stratification. *Ann. Intern. Med.* **2003**, *139*, 137–147. [CrossRef]

12. KDIGO working group. KDIGO 2012 clinical practice guidelines for evaluation and management of chronic kidney disease. *Kidney Int.* **2013**, *3*, 1–150.

13. Thomas, M.E.; Blaine, C.; Dawnay, A.; Devonald, M.A.; Ftouh, S.; Laing, C.; Latchem, S.; Lewington, A.; Milford, D.V.; Ostermann, M. The definition of acute kidney injury and its use in practice. *Kidney Int.* **2015**, *87*, 62–73. [CrossRef]

14. Shin, S.R.; Kim, W.H.; Kim, D.J.; Shin, I.W.; Sohn, J.T. Prediction and Prevention of Acute Kidney Injury after Cardiac Surgery. *Biomed. Res. Int.* **2016**, *2016*, 2985148. [CrossRef] [PubMed]

15. Govindarajulu, U.S.; Spiegelman, D.; Thurston, S.W.; Ganguli, B.; Eisen, E.A. Comparing smoothing techniques in Cox models for exposure-response relationships. *Stat. Med.* **2007**, *26*, 3735–3752. [CrossRef] [PubMed]

16. Durrleman, S.; Simon, R. Flexible regression models with cubic splines. *Stat. Med.* **1989**, *8*, 551–561. [CrossRef] [PubMed]

17. Demler, O.V.; Paynter, N.P.; Cook, N.R. Tests of calibration and goodness-of-fit in the survival setting. *Stat. Med.* **2015**, *34*, 1659–1680. [CrossRef]

18. Harrell, F.E., Jr.; Lee, K.L.; Mark, D.B. Multivariable prognostic models: Issues in developing models, evaluating assumptions and adequacy, and measuring and reducing errors. *Stat. Med.* **1996**, *15*, 361–387. [CrossRef]

19. Yokoyama, M.; Fujii, Y.; Iimura, Y.; Saito, K.; Koga, F.; Masuda, H.; Kawakami, S.; Kihara, K. Longitudinal change in renal function after radical nephrectomy in Japanese patients with renal cortical tumors. *J. Urol.* **2011**, *185*, 2066–2071. [CrossRef]

20. Barlow, L.J.; Korets, R.; Laudano, M.; Benson, M.; McKiernan, J. Predicting renal functional outcomes after surgery for renal cortical tumours: A multifactorial analysis. *BJU Int.* **2010**, *106*, 489–492. [CrossRef]

21. Huang, W.C.; Levey, A.S.; Serio, A.M.; Snyder, M.; Vickers, A.J.; Raj, G.V.; Scardino, P.T.; Russo, P. Chronic kidney disease after nephrectomy in patients with renal cortical tumours: A retrospective cohort study. *Lancet Oncol.* **2006**, *7*, 735–740. [CrossRef]

22. Li, L.; Lau, W.L.; Rhee, C.M.; Harley, K.; Kovesdy, C.P.; Sim, J.J.; Jacobsen, S.; Chang, A.; Landman, J.; Kalantar-Zadeh, K. Risk of chronic kidney disease after cancer nephrectomy. *Nat. Rev. Nephrol.* **2014**, *10*, 135–145. [CrossRef] [PubMed]

23. Patel, H.D.; Pierorazio, P.M.; Johnson, M.H.; Sharma, R.; Iyoha, E.; Allaf, M.E.; Bass, E.B.; Sozio, S.M. Renal Functional Outcomes after Surgery, Ablation, and Active Surveillance of Localized Renal Tumors: A Systematic Review and Meta-Analysis. *Clin. J. Am. Soc. Nephrol.* **2017**, *12*, 1057–1069. [CrossRef] [PubMed]

24. McIntosh, A.G.; Parker, D.C.; Egleston, B.L.; Uzzo, R.G.; Haseebuddin, M.; Joshi, S.S.; Viterbo, R.; Greenberg, R.E.; Chen, D.Y.T.; Smaldone, M.C.; et al. Prediction of significant estimated glomerular filtration rate decline after renal unit removal to aid in the clinical choice between radical and partial nephrectomy in patients with a renal mass and normal renal function. *BJU Int.* **2019**, *124*, 999–1005. [CrossRef]

25. Fergany, A.F.; Hafez, K.S.; Novick, A.C. Long-term results of nephron sparing surgery for localized renal cell carcinoma: 10-year followup. *J. Urol.* **2000**, *163*, 442–445. [CrossRef]
26. Patard, J.J.; Shvarts, O.; Lam, J.S.; Pantuck, A.J.; Kim, H.L.; Ficarra, V.; Cindolo, L.; Han, K.R.; De La Taille, A.; Tostain, J.; et al. Safety and efficacy of partial nephrectomy for all T1 tumors based on an international multicenter experience. *J. Urol.* **2004**, *171*, 2181–2185. [CrossRef]
27. Klarenbach, S.; Moore, R.B.; Chapman, D.W.; Dong, J.; Braam, B. Adverse renal outcomes in subjects undergoing nephrectomy for renal tumors: A population-based analysis. *Eur. Urol.* **2011**, *59*, 333–339. [CrossRef]
28. Sun, M.; Bianchi, M.; Hansen, J.; Trinh, Q.D.; Abdollah, F.; Tian, Z.; Sammon, J.; Shariat, S.F.; Graefen, M.; Montorsi, F.; et al. Chronic kidney disease after nephrectomy in patients with small renal masses: A retrospective observational analysis. *Eur. Urol.* **2012**, *62*, 696–703. [CrossRef]
29. Alam, R.; Patel, H.D.; Osumah, T.; Srivastava, A.; Gorin, M.A.; Johnson, M.H.; Trock, B.J.; Chang, P.; Wagner, A.A.; McKiernan, J.M.; et al. Comparative effectiveness of management options for patients with small renal masses: A prospective cohort study. *BJU Int.* **2019**, *123*, 42–50. [CrossRef]
30. Campbell, S.C.; Novick, A.C.; Belldegrun, A.; Blute, M.L.; Chow, G.K.; Derweesh, I.H.; Faraday, M.M.; Kaouk, J.H.; Leveillee, R.J.; Matin, S.F.; et al. Guideline for management of the clinical T1 renal mass. *J. Urol.* **2009**, *182*, 1271–1279. [CrossRef]
31. Coca, S.G.; Singanamala, S.; Parikh, C.R. Chronic kidney disease after acute kidney injury: A systematic review and meta-analysis. *Kidney Int.* **2012**, *81*, 442–448. [CrossRef] [PubMed]
32. Ishani, A.; Xue, J.L.; Himmelfarb, J.; Eggers, P.W.; Kimmel, P.L.; Molitoris, B.A.; Collins, A.J. Acute kidney injury increases risk of ESRD among elderly. *J. Am. Soc. Nephrol.* **2009**, *20*, 223–228. [CrossRef]
33. Lo, L.J.; Go, A.S.; Chertow, G.M.; McCulloch, C.E.; Fan, D.; Ordonez, J.D.; Hsu, C.Y. Dialysis-requiring acute renal failure increases the risk of progressive chronic kidney disease. *Kidney Int.* **2009**, *76*, 893–899. [CrossRef] [PubMed]
34. Kim, J.S.; Kim, Y.J.; Ryoo, S.M.; Sohn, C.H.; Seo, D.W.; Ahn, S.; Lim, K.S.; Kim, W.Y. One—Year Progression and Risk Factors for the Development of Chronic Kidney Disease in Septic Shock Patients with Acute Kidney Injury: A Single-Centre Retrospective Cohort Study. *J. Clin. Med.* **2018**, *7*, 554. [CrossRef] [PubMed]
35. Hori, D.; Katz, N.M.; Fine, D.M.; Ono, M.; Barodka, V.M.; Lester, L.C.; Yenokyan, G.; Hogue, C.W. Defining oliguria during cardiopulmonary bypass and its relationship with cardiac surgery-associated acute kidney injury. *Br. J. Anaesth.* **2016**, *117*, 733–740. [CrossRef]
36. Mizota, T.; Yamamoto, Y.; Hamada, M.; Matsukawa, S.; Shimizu, S.; Kai, S. Intraoperative oliguria predicts acute kidney injury after major abdominal surgery. *Br. J. Anaesth.* **2017**, *119*, 1127–1134. [CrossRef]
37. Gaffney, A.M.; Sladen, R.N. Acute kidney injury in cardiac surgery. *Curr. Opin. Anaesthesiol.* **2015**, *28*, 50–59. [CrossRef]
38. Anderson, R.J.; Linas, S.L.; Berns, A.S.; Henrich, W.L.; Miller, T.R.; Gabow, P.A.; Schrier, R.W. Nonoliguric acute renal failure. *N. Engl. J. Med.* **1977**, *296*, 1134–1138. [CrossRef]
39. Anderson, R.G.; Bueschen, A.J.; Lloyd, L.K.; Dubovsky, E.V.; Burns, J.R. Short-term and long-term changes in renal function after donor nephrectomy. *J. Urol.* **1991**, *145*, 11–13. [CrossRef]
40. Vincenti, F.; Amend, W.J., Jr.; Kaysen, G.; Feduska, N.; Birnbaum, J.; Duca, R.; Salvatierra, O. Long-term renal function in kidney donors. Sustained compensatory hyperfiltration with no adverse effects. *Transplantation* **1983**, *36*, 626–629. [CrossRef]
41. Talseth, T.; Fauchald, P.; Skrede, S.; Djoseland, O.; Berg, K.J.; Stenstrom, J.; Heilo, A.; Brodwall, E.K.; Flatmark, A. Long-term blood pressure and renal function in kidney donors. *Kidney Int.* **1986**, *29*, 1072–1076. [CrossRef] [PubMed]
42. Najarian, J.S.; Chavers, B.M.; McHugh, L.E.; Matas, A.J. 20 years or more of follow-up of living kidney donors. *Lancet* **1992**, *340*, 807–810. [CrossRef]
43. Giglio, M.; Dalfino, L.; Puntillo, F.; Brienza, N. Hemodynamic goal-directed therapy and postoperative kidney injury: An updated meta-analysis with trial sequential analysis. *Crit. Care* **2019**, *23*, 232. [CrossRef] [PubMed]
44. Hur, M.; Park, S.K.; Yoo, S.; Choi, S.N.; Jeong, C.W.; Kim, W.H.; Kim, J.T.; Kwak, C.; Bahk, J.H. The association between intraoperative urine output and postoperative acute kidney injury differs between partial and radical nephrectomy. *Sci. Rep.* **2019**, *9*, 1–9. [CrossRef] [PubMed]

45. Jeon, H.G.; Jeong, I.G.; Lee, J.W.; Lee, S.E.; Lee, E. Prognostic factors for chronic kidney disease after curative surgery in patients with small renal tumors. *Urology* **2009**, *74*, 1064–1068. [CrossRef] [PubMed]
46. Kim, N.Y.; Hong, J.H.; Koh, D.H.; Lee, J.; Nam, H.J.; Kim, S.Y. Effect of Diabetes Mellitus on Acute Kidney Injury after Minimally Invasive Partial Nephrectomy: A Case-Matched Retrospective Analysis. *J. Clin. Med.* **2019**, *8*, 468. [CrossRef]
47. Bijol, V.; Mendez, G.P.; Hurwitz, S.; Rennke, H.G.; Nose, V. Evaluation of the nonneoplastic pathology in tumor nephrectomy specimens: Predicting the risk of progressive renal failure. *Am. J. Surg. Pathol.* **2006**, *30*, 575–584. [CrossRef]
48. Kil, H.K.; Kim, J.Y.; Choi, Y.D.; Lee, H.S.; Kim, T.K.; Kim, J.E. Effect of Combined Treatment of Ketorolac and Remote Ischemic Preconditioning on Renal Ischemia-Reperfusion Injury in Patients Undergoing Partial Nephrectomy: Pilot Study. *J. Clin. Med.* **2018**, *7*, 470. [CrossRef]
49. Bravi, C.A.; Larcher, A.; Capitanio, U.; Mari, A.; Antonelli, A.; Artibani, W.; Barale, M.; Bertini, R.; Bove, P.; Brunocilla, E.; et al. Perioperative Outcomes of Open, Laparoscopic, and Robotic Partial Nephrectomy: A Prospective Multicenter Observational Study (The RECORd 2 Project). *Eur. Urol. Focus* **2019**. [CrossRef]

Journal of *Clinical Medicine*

Review

GLP-1 Receptor Agonists and Diabetic Kidney Disease: A Call of Attention to Nephrologists

José Luis Górriz [1,*,†], María José Soler [2,†,‡], Juan F. Navarro-González [3,†], Clara García-Carro [2,†,‡], María Jesús Puchades [1,†], Luis D'Marco [1,†], Alberto Martínez Castelao [4,†,‡], Beatriz Fernández-Fernández [5,†], Alberto Ortiz [4,†,‡], Carmen Górriz-Zambrano [6], Jorge Navarro-Pérez [7] and Juan José Gorgojo-Martinez [8]

1. Nephrology Department, Hospital Clínico Universitario, INCLIVA, Universidad de Valencia, 46010 Valencia, Spain; chuspuchades@gmail.com (M.J.P.); luisgerardodg@hotmail.com (L.D.)
2. Nephrology Department, Hospital Universitari Vall d'Hebron, Universitat Autònoma de Barcelona, 08035 Barcelona, Spain; mjsoler01@gmail.com (M.J.S.); clara.garcia@vhebron.net (C.G.-C.)
3. Unidad de Investigación y Servicio de Nefrología, Hospital Universitario Nuestra Señora de Candelaria, Santa Cruz de Tenerife, Universidad de La Laguna, 38200 Tenerife, Spain; jnavgon@gobiernodecanarias.org
4. IIS-Fundación Jimenez Diaz UAM and School of Medicine, Universidad Autonoma de Madrid, 28040 Madrid, Spain; albertomcastelao@gmail.com (A.M.C.); aortiz@fjd.es (A.O.)
5. Nephrology Department, Bellvitge University Hospital, Hospitalet, 08907 Barcelona, Spain; beaff26@hotmail.com
6. CAP Sant Pere, ABS Reus 1, 43202 Tarragona, Spain; carmengorrizz@gmail.com
7. Hospital Clínico Universitario Valencia, INCLIVA, Universidad de Valencia, 46010 Valencia, Spain; jorgenavper@gmail.com
8. Unidad de Endocrinología y Nutrición, Fundación Hospital Alcorcón, 28922 Madrid, Spain; juanjo.gorgojo@gmail.com
* Correspondence: jlgorriz@senefro.org; Tel.: +34-961973811
† Author is from GEENDIAB (Grupo Español de Estudio de la Nefropatía Diabética).
‡ Author is from REDINREN (Red de investigación Renal, Instituto de Salud Carlos III).

Received: 25 February 2020; Accepted: 26 March 2020; Published: 30 March 2020

Abstract: Type 2 diabetes mellitus (T2DM) represents the main cause of chronic kidney disease (CKD) and end-stage renal disease (ESKD), and diabetic kidney disease (DKD) is a major cause of morbidity and mortality in diabetes. Despite advances in the nephroprotective treatment of T2DM, DKD remains the most common complication, driving the need for renal replacement therapies (RRT) worldwide, and its incidence is increasing. Until recently, prevention of DKD progression was based around strict blood pressure (BP) control, using renin–angiotensin system blockers that simultaneously reduce BP and proteinuria, adequate glycemic control and control of cardiovascular risk factors. Glucagon-like peptide-1 receptor agonists (GLP-1RA) are a new class of anti-hyperglycemic drugs shown to improve cardiovascular and renal events in DKD. In this regard, GLP-1RA offer the potential for adequate glycemic control in multiple stages of DKD without an increased risk of hypoglycemia, preventing the onset of macroalbuminuria and slowing the decline of glomerular filtration rate (GFR) in diabetic patients, also bringing additional benefit in weight reduction, cardiovascular and other kidney outcomes. Results from ongoing trials are pending to assess the impact of GLP-1RA treatments on primary kidney endpoints in DKD.

Keywords: chronic kidney disease; diabetic kidney disease; GLP-1

---

1. Introduction

Chronic kidney disease (CKD) is among the most common complications of type 2 diabetes mellitus (T2DM). In several contemporary studies, 28% to 43% of patients with T2DM have CKD, defined as either a glomerular filtration rate (GFR) below 60 mL/min/1.73 m^2 or a urinary albumin to creatinine ratio (UACR) >30 mg/g [1,2]. In pivotal studies such as UKPDS (UK Prospective Diabetes Study), 25% of T2DM patients developed microalbuminuria within 10 years of diagnosis. Relatively fewer patients develop macroalbuminuria, but in those who do, the death rate is higher than the rate of progression to more advanced nephropathy. Of note, the low rate of renal replacement therapy (RRT) observed likely reflects the high mortality of patients with diabetes-associated CKD, which may reach up to 20% per year [3,4].

Diabetic kidney disease (DKD) is a leading cause of morbidity and mortality in diabetes. CKD is associated with all-cause and cardiovascular mortality in patients with diabetes. Indeed, the excess mortality of diabetes occurs mainly in individuals with diabetes and proteinuria and stems not only from end-stage renal disease (ESRD) but also from cardiovascular disease. As such, DKD is on track to surpass all other diabetic complications in terms of attributable morbidity and mortality [4]. Indeed, on a global level, CKD is expected to become the fifth cause of death worldwide within 20 years, and diabetes-associated CKD is a key contributor to the increasing prevalence of CKD [5].

In recent years it has become apparent that diabetes-associated CKD is more heterogeneous than previously thought, and may consist predominantly of tubular, interstitial and/or vascular involvement rather than glomerular injury. Thus, patients with decreased GFR may not have pathological albuminuria, and the term DKD is now preferred, leaving the term diabetic nephropathy for histologically confirmed lesions [6]. Despite advances in the nephroprotective treatment of T2DM, DKD remains the most common complication, and its incidence keeps increasing, driving the need for RRT worldwide [6,7]. It is, therefore, vital to optimize risk stratification and therapeutic strategies for both DM2 and DKD patients in order to decrease associated morbimortality, mainly from cardiovascular disease, and the need for RRT. Our review will focus on the potential role of glucagon-like peptide-1 receptor agonists (GLP-1RA) as nephroprotective agents in T2DM.

2. Management of Diabetic Kidney Disease

Until recently, prevention of DKD progression was based on strict blood pressure (BP) control, using renin–angiotensin system blockers that simultaneously reduce BP and proteinuria, adequate glycemic control and control of cardiovascular risk factors (dyslipidemia, obesity and smoking), as well as nephrotoxic drug avoidance [6,8,9] (Table 1). This approach nonetheless results in significant residual renal and cardiovascular risk.

Table 1. Risk factors for diabetic kidney disease.

Risk Factor	Susceptibility	Initiation	Progression
Demographic			
Older age	+		
Sex (men)	+		
Race (black, other ethnic minorities)	+		+
Reduced renal mass	+		+
Low birth weight	+		
Low socioeconomic level		+	+
Hereditary			
Family history of DKD	+		
Genetic kidney disease		+	
Systemic conditions			
Hyperglycemia (poorly controlled)	+	+	+
Obesity	+	+	+
Hypertension (poorly controlled)	+		+
Kidney injuries			

Table 1. Cont.

Risk Factor	Susceptibility	Initiation	Progression
Acute kidney injury		+	+
Toxins, nephrotoxic drugs, mainly NSAIDs		+	+
Smoking			+
Urological problems (infection, obstruction)		+	+
Dietary factors			
High protein intake	+		+

DKD: diabetic kidney disease; NSAIDs: nonsteroidal anti-inflammatory drugs. Modified from Alicic [6].

2.1. BP Control

BP should be controlled with drugs that reduce cardiovascular events (angiotensin-converting enzyme inhibitors, angiotensin II receptor blockers, thiazide-like diuretics or calcium channel blockers) [8]. However, despite the controversy surrounding target BP levels in T2DM patients, over 50% have BP <140/90 mmHg, but only around 20% have BP <130/80 mmHg [9].

2.2. Blood Glucose Control

Optimizing glucose control in DKD patients is challenging due to an increased risk of adverse effects from certain glucose-lowering agents. Both UKPDS and DCCT have clearly demonstrated that progression of retinopathy and nephropathy is linked to glycemic control and it is crucial that patients maintain HbA1c levels less than or equal to 6.5% to minimize disease progression [10]. Nevertheless, the drugs previously available to reach this goal were associated with a high risk of hypoglycemia. In the last few decades, therefore, blood glucose objectives have become more moderate, which has undoubtedly had an impact on microvascular complications, including nephropathy. Accumulation of kidney-excreted drugs, especially sulfonylureas and insulin, may result in repeated hypoglycemic episodes [11], while metformin may cause potentially lethal lactic acidosis [12,13]. Some drugs that can be administered despite low eGFR values are marred by side effects such as hypoglycemia with repaglinide, volume retention and risk of distal fractures with pioglitazone, or may impact negatively on CKD progression (e.g., dipeptidyl peptidase 4 inhibitors).

Among the agents used to treat hyperglycemia, only sodium–glucose co-transporter type 2 inhibitors (SGLT2i) and certain GLP-1RA have been shown to improve kidney outcomes regardless of glycemic control [14]. SGLT2i decreased both primary and secondary kidney endpoints in patients with DKD or at high cardiovascular risk, respectively [15–17]. These endpoints include progressive loss of GFR, new-onset albuminuria and RRT initiation, alone or in combination. However, the use of SGLT2i is still limited by the high cost and regulatory contraindications for patients with eGFR <60 mL/min/1.73 m^2. It should be noted that this limit is driven by the perceived lack of benefit on glucose control at these low eGFR levels, since the mechanism of antihyperglycemic action depends on glycosuria. Nevertheless, recent trials demonstrated nephroprotection in patients with GFR >30 mL/min/1.73 m^2, even when the drug was maintained until the initiation of dialysis [16,18]. Some GLP-1RA, including liraglutide [19], semaglutide [20] and dulaglutide [21], have also shown renal benefits, and contrary to SGLT2i, can currently be administered up to an eGFR of 15 mL/min/1.73 m^2.

2.3. Obesity

The prevalence of obesity and abdominal obesity in T2DM patients remains high. In Spain, the country projected to have the longest life expectancy by 2040, this prevalence is 50% and 68%, respectively [5,22]. Obesity and overweight are known renal progression factors, although most studies on the renoprotective impact of weight loss are retrospective or are prospective but with only one-arm observational studies. In obese patients (with and without diabetes), weight loss induces a significant reduction in proteinuria [23], which is observed rapidly after weight loss and correlates with weight reduction. Similarly, the correction of metabolic syndrome components along with weight reduction has been associated with slower DKD progression [24].

In T2DM patients with obesity and advanced CKD (GFR < 40 mL/min/1.73 m^2 and albuminuria > 30 mg/g), the combination of a hypocaloric diet and physical exercise was associated with 12% weight and 36% albuminuria reduction, accompanied by improved renal function and glycemic control [25]. This benefit was recently demonstrated in the SOS study (Swedish Obese Subjects). This study compared 2010 patients undergoing bariatric surgery (17% with T2DM) with 2037 obese controls (13% with T2DM) who received standard care for obesity and an average follow-up of 18 years. Patients undergoing bariatric surgery had a 67% reduction in the incidence of renal endpoint (CKD stage 4–5, dialysis or transplant) compared to controls ($p < 0.001$). The beneficial impact was more evident in patients with pathological albuminuria [26].

3. GLP-1 Receptor Agonists

The incretin effect describes the phenomenon whereby, in healthy individuals, oral glucose elicits higher insulin secretory responses than intravenous glucose, despite inducing similar levels of glycemia. This effect, which is uniformly defective in T2DM patients, is mediated by the gut-derived incretin hormones glucose-dependent insulinotropic polypeptide (GIP) and glucagon-like peptide-1 (GLP-1). Two incretin hormones have been identified produced by entero-endocrine K cells, whereas GLP-1 is mainly secreted from L cells located throughout the intestine, more abundantly towards the distal ileum and colon. GLP-1 acts through binding to its receptor (GLP-1R) triggering a downstream signaling cascade. GLP-1R is a class B G protein-coupled receptor not only expressed in the pancreas and central nervous system, but also detected in lower levels in the gut, kidneys, lungs, liver, heart, muscle, peripheral nervous system and other tissues. GLP-1 increases insulin secretion in response to nutrients, particularly glucose (the so-called incretin effect) and suppresses glucagon secretion from pancreatic islet cells, with a reduction in postprandial glucose levels as the net result [27].

GLP-1 seems also to play a role in the central regulation of feeding by increasing satiety signals and reducing appetite, resulting in decreased food intake and subsequent weight loss. Further, GLP-1 exerts effects on the gastrointestinal tract by slowing the gastric emptying rate and small intestine peristalsis, which conditions slower absorption of glucose. Depending on the molecule (native or recombinant long-acting) and administration route, GLP-1RA have broad pleiotropic action on metabolism. All these actions are transduced by a single GLP-1R located in many organs including the kidney. Among the numerous beneficial effects mediated by GLP-1RA are blood glucose regulation, body weight reduction due to food intake inhibition and reduced gastric motility, cell proliferation stimulation, inflammation and apoptosis reduction, and improved cardiovascular function, neuroprotection and renoprotection [28,29].

In T2DM individuals, circulating GLP-1 levels are similar to those found in normoglycemic individuals, yet partial resistance to the insulinotropic effects of GLP-1RA is seen in some T2DM patients at physiological and pharmacological concentrations.

GLP-1 response to oral glucose tolerance test is up to 25% lower in individuals with prediabetes or T2DM than in those with normal glucose regulation. Whether a defective incretin system in T2DM is caused by decreased responsiveness of β cells to GLP-1 and glucose-dependent insulinotropic polypeptide (GIP), or by hyposecretion of incretin hormones, remains unclear. Importantly, in T2DM the insulinotropic response to exogenous GIP administration is completely lost, while a partially preserved, substantial dose-dependent response to GLP-1 is observed [30].

Diminished insulin secretion in response to treatment with GLP-1RA has been associated with genetic and metabolic alterations, thus implicating genetic variation in the transcription factor 7-like 2 (TCF7L2), the loci for GLP-1R, wolfram syndrome 1 and chymotrypsinogen B1/2 [31]. Impaired proinsulin conversion could explain the mechanism for TCF7L2-associated diminished GLP-1RA efficacy as well as dependent repression of GLP-1R expression on b-cells [32]. Incretin action may also be reduced during hyperglycemia and also in some individuals with prediabetes, diabetes and insulin resistance [28]. Native GLP-1 has a very short half-life (about two minutes) because of rapid degradation by the endogenous enzymes dipeptidyl peptidase (DPP-4) and neutral endopeptidase.

GLP-1RA are a pharmacological family of peptides that stimulate the human GLP-1 receptor. They can be classified according to different characteristics, such as molecular size, chemical structure and duration of action. Based on their chemical structure, GLP-1 RA can be divided into two groups: incretin-mimetics (exendin-4 analogs) and human GLP-1RA. Incretin-mimetics are derived from exendin-4, a 39 amino acid peptide isolated from the saliva of the giant lizard Gila monster (Heloderma suspectum), and they share a 53% homology with human GLP-1. Such structural differences with human GLP-1 confer both exendin-4 and incretin-mimetics with resistance to inactivation by DPP-4. Currently, daily exenatide, once-weekly exenatide and lixisenatide are approved exendin-4 analogs, and efpeglenatide is an investigational drug in Phase 3 clinical development. Incretin-mimetic drugs are mainly eliminated by glomerular filtration, tubular reabsorption and subsequent proteolytic degradation, so their clearance is reduced in patients with renal insufficiency. Due to their partial homology to human GLP-1, these drugs may induce an immune response in some individuals; approximately 2–3% of patients develop inactivating antibodies. Human GLP-1 analogs show a close structural homology to native GLP-1, and they increase their half-life through several molecular modifications, such as attachment of fatty acid (liraglutide and semaglutide), albumin (albiglutide) or a constant fraction of immunoglobulin G4 (dulaglutide). Single substitutions of amino acids in the molecular sequence of native GLP-1 confer human GLP-1RA resistance to DPP-4 cleavage. Liraglutide, dulaglutide and semaglutide (subcutaneous and oral) are currently approved human GLP-1RA, since albiglutide was withdrawn from the market for commercial reasons. Human GLP-1RA are metabolized in target tissues via the common proteolytic pathway of large proteins, without a specific organ identified as the main route of elimination. They are protected from renal clearance by either their large molecular size or their noncovalent attachment to albumin. Due to their structural similarity to native human GLP-1, their immunogenicity is low [33].

Large-sized GLP-1RA, such as dulaglutide and albiglutide, are bound to large proteins, which makes it difficult for them to cross the blood–brain barrier and reach satiety centers. Therefore, a lower effect on body weight has been seen in clinical trials with these molecules in comparison to small-sized GLP-1 RA such as liraglutide or semaglutide [33].

Classification of GLP-1 RA into short- and long-acting agonists may be more useful from a practical viewpoint. Long-acting GLP-1 RA, such as liraglutide, dulaglutide, once-weekly exenatide or semaglutide, induce prolonged stimulation of the GLP-1R, leading to a greater reduction of fasting plasma glucose and HbA1c. Short-acting GLP-1 RA, such as daily exenatide or lixisenatide, show a lower effect on fasting plasma glucose, basal insulin secretion and HbA1c but induce a more pronounced decrease in postprandial glycemia. The effect of short-acting GLP1-RA on postprandial glycemia seems to be a consequence of delayed gastric emptying. In contrast, the effect of long-acting GLP-1 RA on gastric emptying is lost after a few weeks of treatment due to a tachyphylaxis mechanism [33].

Human GLP-1RA (liraglutide, semaglutide, albiglutide, dulaglutide) have been shown to reduce cardiovascular morbidity and mortality. In contrast, incretin-mimetics (daily exenatide, once-weekly exenatide, lixisenatide) have not demonstrated superiority in cardiovascular outcomes trials. Some experts argue that differences in study design and population, statistical power, adherence, withdrawals or even mere chance may explain these findings [33]. However, certain molecular differences will probably have contributed, at least partially, to the results:

- Immunogenicity. Exendin-4 analogs are immunogenic, and, in some patients, the drug is inactivated by antibodies, whereas human GLP-1RA rarely induce antibody formation.
- Tolerance. Exendin-4 analogs are metabolized and eliminated by the kidneys, so they accumulate in CKD patients (precisely those with the highest cardiovascular risk), which may favor their withdrawal due to gastrointestinal intolerance. Human GLP-1RA, in contrast, are not eliminated by the kidneys.
- Production of GLP-$_{19\text{-}36}$. Exendin-4 analogs are fully resistant to inactivation by DPP-4. Conversely, human GLP-1RA may be partially metabolized to small amounts of the metabolite GLP-$_{19\text{-}36}$, which could have an additional cardioprotective effect, acting in the endothelial mitochondria through a non-GLP-1 receptor pathway.

From a pharmacological point of view, GLP-1RA effects are mediated by their target molecule (GLP-1R). In this regard, Takayanagi et al. reported that target molecular occupancy could be a useful parameter for evaluating the clinical efficacy of these drugs [34]. They showed that GLP-1RA produce their clinical effect at a relatively low level of GLP-1R occupancy, suggesting that this parameter (GLP-1R occupancy rate) could be used to evaluate clinical efficacy irrespective of the drugs used. Moreover, applied to a single patient, it would be possible to evaluate the clinical efficacy of these drugs individually to make optimal treatment choices.

4. GLP-1RA in Diabetic Kidney Disease Patients

4.1. Pharmacokinetics

The structural difference in human GLP-1 confers exendin-4 and, by extension, incretin-mimetics, resistance to inactivation by DPP-4. Incretin-mimetics are eliminated mainly by glomerular filtration, tubular reabsorption and subsequent proteolytic degradation, so their clearance is reduced in patients with renal insufficiency. In contrast, GLP-1 analogs are metabolized locally in the target tissues by the common route of large proteins, without a specific organ identified as the main route of elimination. They are protected from renal clearance by their large molecular size or by their non-covalent binding to albumin [33,35].

A pharmacokinetic study with the administration of a single subcutaneous dose of semaglutide 0.5 mg showed that semaglutide exposure was similar between subjects with mild/moderate renal impairment or end-stage renal disease and subjects with normal renal function. This equivalence was not demonstrated in subjects with severe renal impairment, in which mean exposure was 22% higher [36].

Thus, the European Medicines Agency (EMA) has approved the use of all commercially available human GLP-1 analogs up to eGFR of 15 mL/min/1.73 m^2, while all exendin-4 analogs are contraindicated below 30 mL/min/1.73 m^2, given the risk of accumulation and toxicity [20,37]. GLP-1RA may exert a beneficial action on the kidneys through blood glucose and BP-lowering effects, reduction of insulin levels and weight loss as well as possible direct cardio-nephroprotective mechanisms through actions on endothelial dysfunction and inflammation [38] (Figure 1).

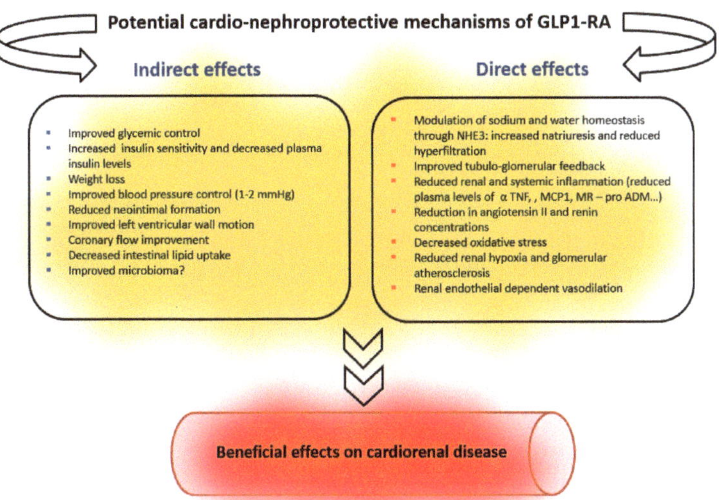

Figure 1. Potential mechanisms for the cardio-nephroprotective effect of GLP-1RA (modified from: Thomas [39] and Greco [40].

4.2. Antihyperglycemic Efficacy of GLP-1RA in Patients with DKD

The range of antihyperglycemic medications in patients with advanced DKD is limited. When GFR falls below 30 mL/min/1.73 m^2, patients are traditionally switched to insulin, metaglinides or DPP4i. Insulin treatment is associated with an increased risk of hypoglycemia [11], weight gain and increased sodium reabsorption in the proximal tubule [41]. DPP4i are modestly effective in reducing glycemia in all ranges of CKD, even in dialysis [42]. However, no cardiovascular or renal benefit has been observed [43], and even saxagliptin was associated with an increased incidence of heart failure [17]. By contrast, SGLT2i and some GLP-1RA were associated with improved cardiorenal outcomes. SGLT2i reduced cardiorenal risk in patients at high cardiovascular risk and in patients with DKD, as recently reported [44,45]. Cardiovascular and renal benefit is present despite only a modest reduction in weight and regardless of glycemic control.

The GLP-1RA liraglutide [19,46], dulaglutide [37] and subcutaneous and oral semaglutide [20,47] have shown to improve glycemic control in patients with diabetes and very low eGFR, including dialysis patients for liraglutide [48]. In T2DM patients with moderate renal impairment, liraglutide demonstrated better glycemic control and weight reduction, with a good safety profile in terms of low hypoglycemia risk compared to placebo [49]. In addition, the AWARD-7 trial was able to demonstrate that when both were combined with insulin lispro, once-weekly dulaglutide attenuated eGFR decline more than titrated daily insulin glargine in T2DM and moderate-to-severe CKD patients. Note that these benefits were observed lowering HbA1c to a similar extent as insulin glargine, but with additional weight loss and the benefit of a lower rate of hypoglycemia in the weekly dulaglutide group as compared with daily insulin glargine, thus demonstrating increased safety [22]. In concordance with these findings, in the PIONEER 5 trial once-daily oral semaglutide 14 mg was superior to placebo in decreasing HbA$_{1c}$ and body weight in patients with T2DM and moderate renal impairment, revealing a good safety profile, including renal safety and few hypoglycemic episodes [50].

5. GLP-1 Receptor Agonists and Cardiovascular Outcomes

The efficacy and safety of GLP-1RA has been evaluated in eight clinical trials with cardiovascular outcomes in 60,090 T2DM patients: ELIXA (lixisenatide) [50], EXSCEL (exenatide) [51], LEADER (liraglutide) [19,52], SUSTAIN 6 (subcutaneous semaglutide) [20], HARMONY (albiglutide) [53], REWIND (dulaglutide) [37], PIONEER 6 (oral semaglutide) [54] and FREEDOM-CVO (ITCA 650, a novel drug device delivering continuous exenatide) [55] (See Table 2, Figure 2).

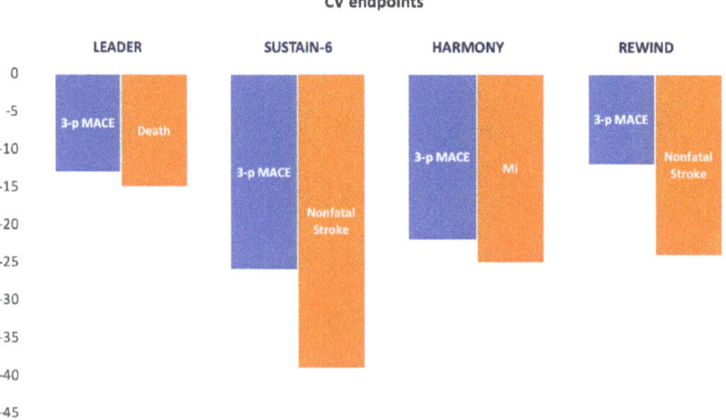

Figure 2. GLP-1RA RCT studies with known beneficial cardiovascular endpoints. Blue bars—percentage of 3-point MACE reduction with the intervention drug. Orange bars—percentage of individual primary cardiovascular endpoint reduction with the intervention drug. Three-point MACE, 3-point MACE; MI, fatal and nonfatal myocardial infarction.

The first reported trial, ELIXA, conducted in 6068 patients with a history of recent acute coronary syndrome and T2DM, was CV neutral, confirming the noninferiority of lixisenatide for three-point MACE [50]. The second one, LEADER, was conducted in 9340 T2DM patients with high cardiovascular risk and demonstrated both CV noninferiority and statistical superiority of once-daily treatment with liraglutide. The reduction in three-point MACE (HR 0.87 (95% CI 0.78–0.97)) with liraglutide was driven mainly by a reduction in CV death (HR 0.78 (95% CI 0.66–0.93)) [52]. The SUSTAIN 6 trial studied the effect of once-weekly treatment with 0.5 or 1 mg of long-acting semaglutide in 3297 T2DM patients at high risk for CV risk, defined by age 50 or older with established cardiovascular disease, chronic heart failure (New York Heart Association class II or III), or CKD stage 3 or higher or age 60 or older with at least one cardiovascular risk factor [20]. This Phase 3 trial demonstrated a favorable effect on three-point MACE accompanied by a significant decrease in nonfatal stroke (HR 0.61 (95% CI 0.38–0.99), p = 0.04). EXSCEL was the largest study performed in a usual-care setting including 14,752 T2DM patients with or without previous CVD. This study was CV neutral, confirming the noninferiority of once-weekly treatment with 2 mg long-acting extended-release exenatide [51]. The HARMONY trial studied the effect of once-weekly treatment with 30 or 50 mg long-acting albiglutide in 9469 T2DM patients with cardiovascular disease [53]. This trial demonstrated a favorable effect on three-point MACE accompanied by a significant decrease in myocardial infarction (HR 0.75 (96% CI 0.61–0.90), p = 0.003) [53]. PIONEER 6 studied the effect of once-daily oral semaglutide treatment with a target dose of 14 mg in 3183 T2DM patients aged 50 or older with established cardiovascular disease or CKD or aged 60 or older with cardiovascular risk factors alone [54]. This trial demonstrated the noninferiority of oral semaglutide for three-point MACE, with a non-significant 21% reduction in MACE in the treatment group [54]. The last published trial, REWIND, was conducted in 9901 T2DM patients with either a previous cardiovascular event or cardiovascular risk factors and demonstrated CV noninferiority and statistical superiority of once-weekly treatment with 1.5 mg dulaglutide. Dulaglutide reduced cardiovascular outcomes in both men and women with or without previous CV disease. The reduction in three-point MACE (HR 0.88 (95% CI 0.79–0.99)) with dulaglutide was mainly driven by a reduction in nonfatal stroke (HR 0.76 (95% CI 0.61–0.95)) [37].

Table 2. Key GLP-1RA RCTs with cardiovascular and renal endpoints.

Drug (Ref)	Trial	*n*	Studied Population	Mean Duration	Composite Primary CV Endpoint	Result HR (95% CI; *p*)	Individual Primary CV Endpoint	Result HR (95% CI; *p*)
Lixisenatide [45]	ELIXA	6068	T2D and acute coronary syndrome	25 m	3P-MACE	Neutral	None	Neutral
Exenatide [46]	EXSCEL	14,752	T2D with or without CVD	3.2 y	3P-MACE	Neutral	None	Neutral
Liraglutide [19,47]	LEADER	9340	T2D and high CV risk	3.8 y	3P-MACE	0.87 (0.78–0.97; *p* < 0.001)	Death from any cause	0.85 (0.74–0.97; *p* = 0.02)
Semaglutide [20] (sc)	SUSTAIN-6	3297	T2D 50 y or more with established CVD, CHF or CKD G3 or higher or >60 y w/CV risk factor	2.1 y	3P-MACE	0.74 (0.58–0.95; *p* = 0.02)	Nonfatal stroke	0.61 (0.38–0.99; *p* = 0.04)
Albiglutide [48]	HARMONY	9469	T2D and CVD or CV risk factors	3.8 y	3P-MACE	0.78 (0.68–0.90; *p* = 0.0006)	Fatal or nonfatal myocardial infarction	0.75 (0.61–0.90, *p* = 0.003)
Dulaglutide [28]	REWIND	9901	T2D and CVD or CV risk factors	5.4 y	3P-MACE	0.88 (0.79–0.99; *p* = 0.026)	Nonfatal Stroke	0.76 (0.61–0.95; *p* = 0.017)
Semaglutide [49] (oral)	PIONEER-6	3183	T2D and CVD or CV risk factors	15.9 m	3P-MACE	Neutral	None	Neutral
Exenatide [22]	FREEDOM-CVO	4000	T2D and CV disease	UK	UK	UK	UK	UK

T2D, type 2 diabetes mellitus; CVD, Cardiovascular disease; 3P-MACE, 3-point MACE (death from cardiovascular causes, nonfatal myocardial infarction, or nonfatal stroke); SC; subcutaneous, UK, unknown; y, years; m, Month; RCTs: randomized clinical trial.

The FREEDOM-CVO trial, designed to test CV safety in a pre-approval setting, evaluated the continuous delivery of exenatide in more than 4000 T2DM patients with CV disease. Although the final results have not yet been published, early on in May 2016, the company press release reported that the study had achieved all its clinical endpoints and was completed on time, confirming noninferiority in terms of CV outcomes [55].

A recently published systematic review and meta-analysis of cardiovascular outcome trials with GLP-1RA screened 27 publications and seven of the above-mentioned clinical trials [56]. Overall, GLP-1 receptor agonist treatment reduced MACE by 12% (HR 0.88, (95% CI 0.82–0.94); $p < 0.0001$). HRs were 0.88 (95% CI 0.81–0.96; $p = 0.003$) for death from cardiovascular causes, 0.84 (0.76–0.93; $p < 0.0001$) for fatal or nonfatal stroke and 0.91 (0.84–1.00; $p = 0.043$) for fatal or nonfatal myocardial infarction. GLP-1 receptor agonist treatment reduced all-cause mortality by 12% (HR 0.88, (95% IC 0.83–0.95; $p = 0.001$)), hospital admission for heart failure by 9% (HR 0.91, (95% IC 0.83–0.99; $p = 0.028$)), without increasing the risk of severe hypoglycemia, pancreatitis or pancreatic cancer [56].

Overall, the clinical trials mentioned in this section demonstrated noninferiority and CV neutrality for exendin-4 GLP-1RA, and cardiovascular protection for human GLP-1RA. This observed differential effect may be ascribed to different causes such as clinical trial design, studied population characteristics, and molecular differences between exendin-4 and human GLP-1RA.

All these data prompt the question of whether GLP-1RA has a class effect. Human GLP-1 analogs (liraglutide, semaglutide, albiglutide, predictably dulaglutide) have demonstrated CV morbidity and mortality reduction. In contrast, exendin-4 or incretin-mimetic analogs (exenatide, exenatide LAR, lixisenatide) have not achieved superiority in CV safety studies, a fact that points towards a possible class effect in this group of drugs. Other explanations for these differences include the study designs, populations studied, potency, withdrawals or even mere chance. However, certain molecular differences are likely to have contributed at least partially to these results, which supports the hypothesis of a class effect in GLP-1RA.

6. GLP-1 Receptor Agonists on Kidney Outcomes

To date, there are no published GLP-1RA trials with a primary endpoint of kidney events or enrolling only DKD patients. For this reason, insights into the renal impact of GLP-1RA were provided by cardiovascular outcomes trials (Table 3, Figure 3).

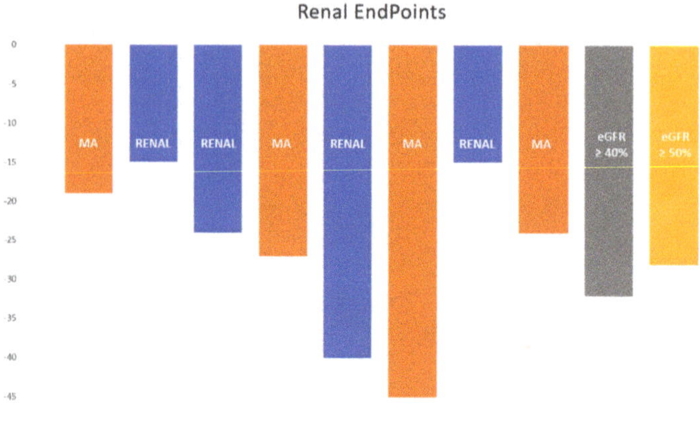

Figure 3. GLP-1RA RCT studies with known beneficial renal endpoints. Blue bars—percentage of composite renal endpoint reduction with the intervention drug. Orange bars—percentage of individual renal endpoint reduction with the intervention drug. Renal, combined renal endpoint; MA, macroalbuminuria; ↓ eGFR ≥ 40%, Sustained decline in eGFR of ≥40%; ↓ eGFR ≥ 50%, Sustained decline in eGFR of ≥50%.

Table 3. Key GLP-1RA RCTs with kidney endpoints.

Drugs	Trials	% n eGFR < 60	Composite Kidney Endpoint	Results	Individual Kidney Endpoint	Result HR (95% CI; p)
Lixisenatide [45]	ELIXA	23	NA	NA	New onset macroalbuminuria	0.808 (0.660–0.991; p = 0.0404)
Exenatide [46]	EXSCEL	17	40% reduction in eGFR loss, onset of dialysis or transplantation, renal death and onset of macroalbuminuria	0.85 (0.73–0.98; p = 0.027)	None	Neutral
Liraglutide [19,47]	LEADER	23	New onset macroalbuminuria, sustained serum creatinine duplication, initiation of renal replacement therapy or renal death	0.78 (0.67–0.92; p = 0.003)	New onset macroalbuminuria	0.74 (0.37–0.77; p = 0.001)
Semaglutide [20] (sc)	SUSTAIN-6	28.5	New onset macroalbuminuria, doubling serum creatinine reaching an eGFR <45 mL/min/1.73 m^2, initiation of renal replacement therapy or renal death	0.64 (0.46–0.88; p = 0.005)	Persistent macroalbuminuria	0.54 (0.60–0.91; p = 0.004)
Albiglutide [48]	HARMONY	11	UK	UK	UK	UK
Dulaglutide [28]	REWIND	22	New onset macroalbuminuria, sustained decreased of eGFR <30% or the initiation of renal replacement therapy	0.85 (0.77–0.93, p = 0.0004)	New onset macroalbuminuria; Sustained decline in eGFR of ≥40%; Sustained decline in eGFR of ≥50%	0.77 (0.68–0.87; p < 0.0001); 0.70 (0.57–0.85; p = 0.0004); 0.74 (0.66–0.84; p < 0.0001)
Semaglutide [49] (oral)	PIONEER-6	27	UK	UK	UK	UK
Exenatide [22]	FREEDOM-CVO	UK	UK	UK	UK	UK

NA, not apply; SC; subcutaneous; UK, unknown; RCTs: randomized clinical trial.

In the ELIXA (Evaluation of Lixisenatide in Acute Coronary Syndrome) trial [50,57], eGFR was <60 mL/min/1.73 m^2 in 25% of lixisenatide and in 22% of placebo patients. In an exploratory analysis, lixisenatide decreased the percentage of albuminuria change compared to baseline when tested against placebo, although after adjusting for various variables, this was only significant in patients with macroalbuminuria. However, no differences were observed regarding eGFR loss. In the EXSCEL trial [51], eGFR was <60 mL/min/1.73 m^2 in 17% of exenatide and placebo patients. Secondary renal endpoints were not predefined. However, post-hoc analysis demonstrated a significant reduction in the risk of a combined kidney endpoint consisting of a 40% reduction in the rate of eGFR decline, onset of dialysis or transplantation, renal death and onset of macroalbuminuria (HR 0.85 (95% CI 0.73–0.98, $p = 0.027$)) [58]. In the LEADER trial [19,52], 23% of patients had eGFR <60 mL/min/1.73 m^2, 36% albuminuria >30 mg/g and 10% albuminuria >300 mg/g. Liraglutide decreased by 22% (HR 0.78 (95% CI 0.67–0.92, $p = 0.003$)) the risk of a secondary composite kidney endpoint (new-onset macroalbuminuria, sustained serum creatinine duplication, initiation of RRT or renal death). This benefit was mainly at the expense of macroalbuminuria reduction, without significant benefit in the other composite endpoint parameters. Interestingly, when results were stratified according to the eGFR at baseline, the decrease in eGFR in patients with a baseline estimated GFR of 30 to 59 mL/min/1.73 m^2 was 2 mL/min/1.73 m^2 in the liraglutide group, compared with 4 mL/min/1.73 m^2 in the placebo group (HR 1.07 (95% CI 1.04–1.10, $p < 0.001$)) [52]. In the HARMONY trial [23], eGFR was <60 mL/min/1.73 m^2 in 11% of patients. After 1.6 years of follow-up, no significant benefit of albiglutide was observed in the safety endpoint of the eGFR decline rate. To date, no other kidney endpoints have been analyzed. In the REWIND trial [37], eGFR was <60 mL/min/1.73 m^2 in 22% of patients. Dulaglutide achieved a significant 15% risk reduction (HR 0.85 (95% CI 0.77–0.93, $p = 0.0004$)) in the renal composite endpoint (new-onset macroalbuminuria, sustained decreased of eGFR <30% or the initiation of RRT), mainly driven by decreased new-onset macroalbuminuria. The effect of dulaglutide on renal outcome was further explored in a set of sensitivity analyses, which demonstrated that dulaglutide was associated with reduced incidence of sustained eGFR decline of 40% or more (HR 0.70 (95% CI 0.57–0.85, $p = 0.0004$)), and 50% or more (HR 0.56 (95% CI 0.41–0.76, $p = 0.0002$)). Consistent with these results, a recently published sub-analysis of another antidiabetic class drug, empagliflozin, a sodium glucose cotransporter-2 inhibitor, demonstrated that lower thresholds of eGFR (e.g., 30%) were associated with higher event rates but weaker treatment effect, suggesting that in the future renal endpoints should target an eGFR decline above 40% [59].

In the SUSTAIN 6 trial [20], eGFR < 60 mL/min/1.73 m^2 was present in 28.5% of patients and 12.7% had baseline macroalbuminuria. Semaglutide effected a 36% reduction (HR 0.64 (95% CI 0.46–0.88, $p = 0.005$)) in a secondary combined kidney endpoint (new-onset macroalbuminuria, doubling serum creatinine reaching an eGFR <45 mL/min/1.73 m^2, initiation of renal replacement therapy or renal death). The renal benefit owed mainly to impact on new onset macroalbuminuria (2.7% in the semaglutide group vs. 4.9% in the placebo group, $p = 0.001$). Post-hoc analyses showed a 30% reduction in albuminuria and regression to micro- or normoalbuminuria occurred for all degrees of albuminuria in patients treated with semaglutide [60].

In a post-hoc analysis, patients on 1 mg weekly subcutaneous dose of semaglutide had significantly slower eGFR loss than those on placebo (semaglutide 1 mg: −1.05 [−1.41; −0.69] vs. placebo: −1.92 [−2.18; −1.67] mL/min/1.73 m^2/year, $p < 0.001$). Slower eGFR decline was also observed when patients with eGFR <60 or >60 mL/min/1.73 m^2 were analyzed separately [60]. A recent sub-analysis confirmed a significantly milder eGFR decline with 1 mg semaglutide in patients with baseline eGFR between 30 and 60 mL/min/1.73 m^2 [58]. In the PIONEER 6 trial [54], which evaluated oral semaglutide, 27% of patients had eGFR <60 mL/min/1.73 m^2; however, this study did not include renal endpoints. Interestingly, in REWIND, LEADER and SUSTAIN 6, no differences were observed in RRT components of the combined kidney endpoint.

7. Potential Mechanisms for GLP-1RA-Associated Nephroprotection

GLP-1R have been identified in vascular smooth muscle cells of preglomerular arterioles in human and experimental animal kidneys [61] but cannot be easily detected by immunohistochemistry or imaging, and levels of expression may be below the sensitivity of the techniques [62]. Their presence in renal glomeruli and proximal tubular cells remains a subject of debate [63]. Renal expression of GLP-1R could underlie observations of GLP-1RA nephroprotection [30].

Indirect nephroprotective effects may depend on improvement in conventional risk factors for DKD, such as glycemic control, weight control and BP. Although nephroprotection has been observed with other glucose-lowering drugs that decrease weight such as SGLT2i, the latter possess many other potential nephroprotective mechanisms, while conferring a more modest benefit in weight reduction. Nevertheless, in the LEADER study, the difference in kidney outcome was not altered by adjustment for change in glycemic control, body weight or systolic BP [19].

In a meta-analysis of 26,654 patients from 33 clinical trials including 12,469 patients that received liraglutide (41%) or exenatide (59%), GLP-1RA were associated with a systolic BP reduction of 2.22 mmHg (95% CI: –2.97–1.47). In a separate meta-regression analysis, the degree of systolic BP change was not associated with baseline BP, weight loss or improvement in HbA1c [64]. Reduction in glucose levels and improvements in insulin sensitivity following therapy with GLP-1RA also lead to a decline in insulin levels, with benefits to different organs including the kidneys [39]. Treatment with GLP-1RA is able to modulate the microbiome of mice; accordingly, liraglutide could modulate the composition of the gut microbiota, leading to a more lean-related profile that was consistent with its weight-losing effect [65].

Direct renal and cardiac effects may also contribute to nephroprotection. The potential mechanisms of direct renal benefit of GLP-1RA are multiple and incompletely understood, most having been demonstrated in experimental animals only. These putative renoprotective actions and effects of GLP-1RA on kidneys are shown in Figure 2.

Turning next to the effect on natriuresis, GLP-1 has been demonstrated to induce natriuresis and diuresis, likely involving inhibition of sodium–hydrogen exchanger 3 (NHE3) localized at the brush border of the renal proximal tubular cells [66]. NHE3 activity in the proximal tubule increases distal tubular sodium transport in the kidney to the macula densa, resulting in restored tubular glomerular feedback with a reduction in intraglomerular pressure, hyperfiltration and renin–angiotensin system activation [67]. These findings imply that GLP-1RA are proximal diuretics and renal vasodilators that under healthy conditions only mildly influence tubuloglomerular feedback [68].

GLP-1RA also decrease circulating concentrations of angiotensin II, an effect that also plays a role in the observed increase in renal sodium wasting [29], and which may partially explain the BP-lowering effects of GLP-1RA. Several molecules play a role in glucose metabolism including insulin, ATP and glucose itself, which regulate NHE3 and SGLTs in the kidney, thus suggesting certain indirect natriuretic actions of GLP-1 [32]. Some studies in experimental animals have also suggested the effects of GLP-1 may be mediated by Atrial Natriuretic Peptide [39].

7.1. GLP-1RA and Renal Hemodynamics

The question of whether GLP-1RA modulate glomerular hemodynamics is controversial. GLP-1RA may decrease endothelin-1 and angiotensin II-induced vasoconstriction, thus decreasing glomerular hyperfiltration [63]. However, in healthy overweight men, exenatide infusion was found to acutely induce nitric oxide-dependent glomerular afferent arteriole vasodilation, increasing postprandial GFR by 20% (18–20 mL/min/1.73 m^2, p = 0.021) [69]. Theoretically, however, decreased proximal sodium reabsorption would initiate vasoconstriction of the preglomerular arteriole through tubule-glomerular feedback. Yet, the net effect of exenatide on preglomerular arterioles was vasodilation in the present study, suggesting a stronger direct vasodilation effect. These findings imply that GLP-1RA are proximal diuretics and renal vasodilators that only mildly influence tubule-glomerular feedback under healthy conditions [68].

This discrepancy may be due to the fact that the renal effect of exenatide in healthy men may differ from that in insulin-resistant subjects and T2DM patients, which could explain the above-mentioned decrease in GFR after GLP-1 peptide infusion in these individuals. However, this was not replicated in patients with T2DM [69], which was shown/hypothesized to depend on impaired nitric oxide-dependent vasodilation in T2DM patients. Thus, GLP-1RA have the potential to increase or decrease eGFR, depending on baseline conditions.

The net effect GLP-1RA have on glomerular hemodynamics probably depends on the balance of the potential effect on tubule-glomerular feedback of GLP-1RA induced natriuresis (vasoconstriction of the afferent in some patients) vs. vasodilation by nitric oxide, added to the vasodilator effect of the efferent by inhibition of the renin–angiotensin axis and endothelin. The total sum of these divergent forces determines changes in the eGFR, which could explain why in LEADER [19] and SUSTAIN [20], patients presented different changes in eGFR depending on the subgroups stratified according to eGFR at baseline.

7.2. Modulation of Cyclic Adenosine Monophosphate–Protein Kinase a (Camp/PKA) Signaling and Other Anti-Inflammatory Pathways

GLP-1R activation leads to stimulation of cyclic adenosine monophosphate–protein kinase A pathways, producing antioxidative effects; it, therefore, seems likely that GLP-1 protects the kidney from oxidative injury [40]. In the diabetic nephropathy rat model, GLP-1RA also downregulated expression of several inflammatory biomarkers in rats, such as tubulointerstitial tumor necrosis factor alpha (TNF-α), monocyte chemoattractant protein-1(MCP-1), collagen I, alpha-smooth muscle actin (α-SMA) and fibronectin, all reported to play a role in diabetic nephropathy, as well as ameliorating kidney tubules and tubulointerstitial lesions [70].

Other putative renoprotective actions of GLP-1RA are the improvement of renal hypoxia related to vascular rarefaction induced by hyperglycemia and anti-atherogenic effects, both directly via an effect on glucose, lipids, weight and BP and possibly indirectly via anti-inflammatory and anti-ischemic activity [39].

8. Ongoing Studies and Unanswered Questions

There are three ongoing studies of GLP-1RA in primary kidney outcomes: Effect of GLP-1RA, liraglutide, on DKD (NCT01847313); Effect of LIXIsenatide on the Renal System (ELIXIRS) (NCT02276196); and The FLOW study (Effect of semaglutide versus placebo on the progression of renal impairment in subjects with T2DM and CKD) (NCT03819153) [71]. This last trial is the largest one designed to test the effect of once-weekly treatment with 1 mg long-acting subcutaneous semaglutide in T2DM patients with moderate/advanced CKD and albuminuria. FLOW has recently begun recruiting more than 3000 adult T2DM patients with DKD on maximal tolerated dose RAAS blockers. DKD is defined as either albuminuria 300–5000 mg/g and eGFR 50–75 mL/min/1.73 m^2 or albuminuria 100–5000 mg/g and eGFR 25–50 mL/min/1.73 m^2. The primary composite kidney outcome is composed of a persistent $\geq 50\%$ reduction in eGFR or a persistent eGFR <15 mL/min/1.73 m^2 or the initiation of RRT (dialysis or renal transplant) or renal or cardiovascular death.

Several randomized clinical trial studies with GLP-1RA are currently ongoing. Among them, the Heart Disease Study of Semaglutide in Patients with T2DM (SOUL) (NCT03914326) focused on evaluating the effect of oral semaglutide in preventing cardiovascular events in T2DM patients with cardiovascular disease or CKD. In this clinical trial, the first occurrence of a composite CKD endpoint, namely renal death/onset of persistent 50% or more reduction in eGFR (CKD-EPI)/onset of persistent eGFR (CKD-EPI) below 15 mL/min/1.73 m^2 initiation of chronic renal replacement therapy (dialysis or kidney transplantation) is included in the secondary outcomes. In addition, a new GLP-1RA named efpeglenatide is being tested in T2DM patients with established cardiovascular disease or aged over 50 years (male), 55 years (female) or older with eGFR ≥ 25 and <60 mL/min/1.73 m^2 and at least one cardiovascular risk factor. In the AMPLITUDE-O trial (Effect of Efpeglenatide on Cardiovascular

Outcomes) (NCT03496298), the time to first occurrence of any of the following clinical events: new-onset or progression to macroalbuminuria (>300 mg/g) accompanied by a UACR value increase of $\geq$30% from baseline, sustained a $\geq$40% decrease in eGFR from baseline (for $\geq$30 days), chronic dialysis (for $\geq$90 days), renal transplant, sustained eGFR <15 mL/min/1.73 m^2 (for $\geq$30 days) is also considered as a secondary outcome.

Although several basic research studies and randomized clinical trials focused on GLP-1RA are already underway, the renoprotective mechanisms of GLP-1RA are not completely understood. At the pipeline level, further research is needed on the positive effect in terms of proteinuria, and inhibition of the sodium–hydrogen exchanger 3 (NHE3) localized at the brush border of the renal proximal tubular cells. This direct effect at the proximal renal tubular level may play a role in the differential GFR effect depending on different degrees of DKD. SGLT-2 and GLP-1RA in combination have proven effective in T2DM patients, but whether this drug combination exerts a potentiated or decreased cardiorenal protective effect in patients with moderate-advanced DKD is unknown. The threshold of human GLP-1RA is eGFR under 15 mL/min/1.73 m^2, but this owes mainly to the lack of clinical trials in DKD patients with stage 5 CKD in dialysis programs or kidney transplantation. Thus, the safety of this class of drugs in this specific CKD population is also yet to be resolved. More studies are required in patients with advanced CKD and T2DM.

9. Conclusions

Recently, certain glucose-lowering drugs have shown benefits beyond blood glucose control. SGLT2i and GLP-1RA are associated with improved cardiovascular and kidney outcomes. Currently, SGLT2i has regulatory approval only in patients with eGFR >60 mL/min/1.73 m^2, although this is expected to change soon. In contrast, human GLP-1RA can be used up to eGFR of 15 mL/min/1.73 m^2. GLP-1RA offer the potential for adequate glycemic control in multiple stages of DKD without an increased risk of hypoglycemia and with additional benefits in weight reduction, cardiovascular outcomes and exploratory kidney outcomes. The ongoing FLOW RCT is assessing the impact of semaglutide on primary kidney outcomes in DKD.

Author Contributions: All authors contribute equally on this manuscript. All authors have read and agreed to the published version of the manuscript.

Funding: This research received no external funding.

Acknowledgments: J.F.N.G., A.O.: REDINREN. Instituto de Salud Carlos III (RD16/0009/0022).

Conflicts of Interest: J.L.G. has served as consultant for Boëhringer-Ingelheim, Mundipharma, AstraZeneca, Novonordisk and has received speaker honoraria from Boëhringer-Ingelheim, Mundipharma, AstraZeneca, Novonordisk, Novartis and Eli Lilly. M.J.S. reports personal fees and has served on advisory boards from NovoNordisk, Jansen, Boëhringer-Ingelheim, Mundipharma, AstraZeneca, and Esteve, during the study period. J.F.N.G. has served as a consultant, has received speaker fees or travel support from Abbvie, Amgen, AstraZeneca, Boehringer Ingelheim, Esteve, Genzyme, Lilly, Novartis, Servier, Shire and Vifor Fresenius Medical Care, Renal Pharma. C.G.C. reports speaker honoraria and has served on advisory boards from Astra Zeneca and Boehringer Ingelheim and travel support from Astellas, Menarini, Novartis, Esteve, Sanofi and Novonordisk. MJP declares no conflicts of interest. L.D.M. declares no conflicts of interest. A.M.C. declares speaker honoraria from Boëhringer-Ingelheim, Lilly, Novo-Nordisk, M.S.D. and Esteve, and advisory boards for Boëhringer-Ingelheim, Lilly, M.S.D. B.F.F. reports personal speaker fees from Mundipharma, Esteve and Novartis; has received travel support from Abbvie, Amgen, Menarini, Novartis, Novonordisk and Esteve, and has served as an advisory board consultant from Astrazeneca, Boehringer Ingelheim and Mundipharma outside the submitted work. A.O. has served as a consultant for Mundipharma, Sanofi Genzyme and Freeline and has received speaker honoraria from Amgen, Amicus, Otsuka and Fresenius Medical Care. C.G.Z. declares no conflicts of interest. J.N.P. has served as consultant for Sanofi and declares speaker honoraria from Boëhringer-Ingelheim, AstraZeneca, Eli Lilly and a research grant from Boëhringer-Ingelheim. J.J.G.M. has served as a consultant for AstraZeneca, Janssen Pharmaceuticals, Eli Lilly and Company, Merck Sharp and Dohme, Mundipharma, Novo Nordisk and Pfizer; has given lectures for Abbott, AbbVie Inc, AstraZeneca, Boehringer Ingelheim Pharmaceuticals Inc, Esteve, Janssen Pharmaceuticals, Eli Lilly and Company, Merck Sharp and Dohme, Novo Nordisk, Pfizer, Roche Pharma and Sanofi Aventis, and has conducted research activities for AstraZeneca, Novo Nordisk and Sanofi Aventis.

References

1. Rodriguez-Poncelas, A.; Garre-Olmo, J.; Franch-Nadal, J.; Diez-Espino, J.; Mundet-Tuduri, X.; Barrot-De la Puente, J.; Coll-de Tuero, G.; RedGDPS Study Group. Prevalence of chronic kidney disease in patients with type 2 diabetes in Spain: PERCEDIME2 study. *BMC Nephrol.* **2013**, *14*, 46. [CrossRef] [PubMed]
2. Bailey, R.A.; Wang, Y.; Zhu, V.; Rupnow, M.F.T. Chronic kidney disease in US adults with type 2 diabetes: An updated national estimate of prevalence based on Kidney Disease: Improving Global Outcomes (KDIGO) staging. *BMC Res. Notes* **2014**, *7*, 415. [CrossRef] [PubMed]
3. Adler, A.I.; Stevens, R.J.; Manley, S.E.; Bilous, R.W.; Cull, C.A.; Holman, R.R.; UKPDS GROUP. Development and progression of nephropathy in type 2 diabetes: The United Kingdom Prospective Diabetes Study (UKPDS 64). *Kidney Int.* **2003**, *63*, 225–232. [CrossRef] [PubMed]
4. Afkarian, M.; Sachs, M.C.; Kestenbaum, B.; Hirsch, I.B.; Tuttle, K.R.; Himmelfarb, J.; de Boer, I.H. Kidney disease and increased mortality risk in type 2 diabetes. *J. Am. Soc. Nephrol. JASN* **2013**, *24*, 302–308. [CrossRef]
5. Foreman, K.J.; Marquez, N.; Dolgert, A.; Fukutaki, K.; Fullman, N.; McGaughey, M.; Pletcher, M.A.; Smith, A.E.; Tang, K.; Yuan, C.-W.; et al. Forecasting life expectancy, years of life lost, and all-cause and cause-specific mortality for 250 causes of death: Reference and alternative scenarios for 2016-40 for 195 countries and territories. *Lancet* **2018**, *392*, 2052–2090. [CrossRef]
6. Alicic, R.Z.; Rooney, M.T.; Tuttle, K.R. Diabetic Kidney Disease: Challenges, Progress, and Possibilities. *Clin. J. Am. Soc. Nephrol. CJASN* **2017**, *12*, 2032–2045. [CrossRef]
7. United States Renal Data System. USRDS Annual Data Report Executive Summary USA. 2014. Available online: https://www.usrds.org/2014/view/v2_01.aspx (accessed on 28 March 2020).
8. American Diabetes Association. Cardiovascular Disease and Risk Management: Standards of Medical Care in Diabetes-2019. *Diabetes Care* **2019**, *42* (Suppl. 1), S103–S123. [CrossRef]
9. Martínez-Castelao, A.; Górriz, J.L.; Segura-de la Morena, J.; Cebollada, J.; Escalada, J.; Esmatjes, E.; Fácila, L.; Gamarra, J.; Gràcia, S.; Hernánd-Moreno, J.; et al. Consensus document for the detection and management of chronic kidney disease. *Nefrologia* **2014**, *34*, 243–262.
10. Factors in Development of Diabetic Neuropathy. Baseline analysis of neuropathy in feasibility phase of Diabetes Control and Complications Trial (DCCT). The DCCT Research Group. *Diabetes* **1988**, *37*, 476–481. [CrossRef]
11. Ginsberg, J.S.; Zhan, M.; Diamantidis, C.J.; Woods, C.; Chen, J.; Fink, J.C. Patient-reported and actionable safety events in CKD. *J. Am. Soc. Nephrol.* **2014**, *25*, 1564–1573. [CrossRef]
12. Farrington, K.; Covic, A.; Nistor, I.; Aucella, F.; Clyne, N.; de Vos, L.; Findlay, A.; Fouque, D.; Grodzicki, T.; Iyasere, O.; et al. Clinical Practice Guideline on management of older patients with chronic kidney disease stage 3b or higher (eGFR &alt; 45 mL/min/1.73 m^2): A summary document from the European Renal Best Practice Group. *Nephrol. Dial. Transplant.* **2017**, *32*, 9–16.
13. Martínez-Castelao, A.; Górriz, J.L.; Ortiz, A.; Navarro-González, J.F. ERBP guideline on management of patients with diabetes and chronic kidney disease stage 3B or higher. Metformin for all? *Nefrologia* **2017**, *37*, 567–571. [CrossRef] [PubMed]
14. Górriz, J.; Nieto, J.; Navarro-González, J.; Molina, P.; Martínez-Castelao, A.; Pallardó, L. Nephroprotection by Hypoglycemic Agents: Do We Have Supporting Data? *J. Clin. Med.* **2015**, *4*, 1866–1889. [CrossRef] [PubMed]
15. Wanner, C.; Inzucchi, S.E.; Lachin, J.M.; Fitchett, D.; von Eynatten, M.; Mattheus, M.; Johansen, O.E.; Woerle, H.J.; Broedl, U.C.; Zinman, B.; et al. Empagliflozin and Progression of Kidney Disease in Type 2 Diabetes. *N. Engl. J. Med.* **2016**, *375*, 323–334. [CrossRef] [PubMed]
16. Perkovic, V.; Jardine, M.J.; Neal, B.; Bompoint, S.; Heerspink, H.J.L.; Charytan, D.M.; Edwards, R.; Agarwal, R.; Bakris, G.; Bull, S.; et al. Canagliflozin and renal outcomes in type 2 diabetes and nephropathy. *N. Engl. J. Med.* **2019**, *380*, 2295–2306. [CrossRef] [PubMed]
17. Scirica, B.M.; Bhatt, D.L.; Braunwald, E.; Steg, P.G.; Davidson, J.; Hirshberg, B.; Ohman, P.; Frederich, R.; Wiviott, S.D.; Hoffman, E.B.; et al. Saxagliptin and cardiovascular outcomes in patients with type 2 diabetes mellitus. *N. Engl. J. Med.* **2013**, *369*, 1317–1326. [CrossRef] [PubMed]
18. Wiviott, S.D.; Raz, I.; Bonaca, M.P.; Mosenzon, O.; Kato, E.T.; Cahn, A.; Silverman, M.G.; Zelniker, T.A.; Kuder, J.F.; Murphy, S.A.; et al. Dapagliflozin and Cardiovascular Outcomes in Type 2 Diabetes. *N. Engl. J. Med.* **2019**, *380*, 347–357. [CrossRef]

19. Mann, J.F.E.; Brown-Frandsen, K.; Marso, S.P.; Poulter, N.R.; Rasmussen, S.; Rsted, D.D. Committee and Investigators. Liraglutide and renal outcomes in type 2 diabetes. *N. Engl. J. Med.* **2017**, *31*, 839–848. [CrossRef]

20. Marso, S.P.; Bain, S.C.; Consoli, A.; Eliaschewitz, F.G.; Jódar, E.; Leiter, L.A.; Lingvay, I.; Rosenstock, J.; Seufert, J.; Warren, M.L.; et al. Semaglutide and Cardiovascular Outcomes in Patients with Type 2 Diabetes. *N. Engl. J. Med.* **2016**, *375*, 1834–1844. [CrossRef]

21. Tuttle, K.R.; Lakshmanan, M.C.; Rayner, B.; Busch, R.S.; Zimmermann, A.G.; Woodward, D.B.; Botros, F.T. Dulaglutide versus insulin glargine in patients with type 2 diabetes and moderate-to-severe chronic kidney disease (AWARD-7): A multicentre, open-label, randomised trial. *Lancet Diabetes Endocrinol.* **2018**, *6*, 605–617. [CrossRef]

22. Soriguer, F.; Goday, A.; Bosch-Comas, A.; Bordiú, E.; Calle-Pascual, A.; Carmena, R.; Casamitjana, R.; Castaño, L.; Castell, C.; Catalá, M.; et al. Prevalence of diabetes mellitus and impaired glucose regulation in Spain: The Di@bet.es Study. *Diabetologia* **2012**, *55*, 88–93. [CrossRef] [PubMed]

23. Morales, E.; Praga, M. The effect of weight loss in obesity and chronic kidney disease. *Curr. Hypertens. Rep.* **2012**, *14*, 170–176. [CrossRef] [PubMed]

24. Cuevas-Ramos, D.; Campos-Barrera, E.; Durán-Pérez, E.G.; Almeda-Valdés, P.; Muñoz-Hernández, L.; Gómez-Pérez, F.J. Treatment of metabolic syndrome slows progression of diabetic nephropathy. *Metab. Syndr. Relat. Disord.* **2011**, *9*, 483–489.

25. Friedman, A.N.; Chambers, M.; Kamendulis, L.M.; Temmerman, J. Short-term changes after a weight reduction intervention in advanced diabetic nephropathy. *Clin. J. Am. Soc. Nephrol.* **2013**, *8*, 1892–1898. [CrossRef] [PubMed]

26. Shulman, A.; Peltonen, M.; Sjöström, C.D.; Andersson-Assarsson, J.C.; Taube, M.; Sjöholm, K.; le Roux, C.W.; Carlsson, L.M.; Svensson, P.A. Incidence of end-stage renal disease following bariatric surgery in the Swedish Obese Subjects Study. *Int. J. Obes.* **2018**, *42*, 964–973. [CrossRef] [PubMed]

27. Gautier, J.-F.; Choukem, S.-P.; Girard, J. Physiology of incretins (GIP and GLP-1) and abnormalities in type 2 diabetes. *Diabetes Metab.* **2008**, *34* (Suppl. 2), S65–S72. [CrossRef]

28. Müller, T.D.; Finan, B.; Bloom, S.R.; D'Alessio, D.; Drucker, D.J.; Flatt, P.R.; Fritsche, A.; Gribble, F.; Grill, H.J.; Habener, J.F.; et al. Glucagon-like peptide 1 (GLP-1). *Mol. Metab.* **2019**, *30*, 72–130. [CrossRef]

29. Skov, J. Effects of GLP-1 in the kidney. *Rev. Endocr. Metab. Disord.* **2014**, *15*, 197–207. [CrossRef]

30. Muskiet, M.H.A.; Tonneijck, L.; Smits, M.M.; van Baar, M.J.B.; Kramer, M.H.H.; Hoorn, E.J.; Joles, J.A.; van Raalte, D.H. GLP-1 and the kidney: From physiology to pharmacology and outcomes in diabetes. *Nat. Rev. Nephrol.* **2017**, *13*, 605–628. [CrossRef]

31. Heni, M.; Ketterer, C.; Thamer, C.; Herzberg-Schäfer, S.A.; Guthoff, M.; Stefan, N.; Machicao, F.; Staiger, H.; Fritsche, A.; Häring, H.-U. Glycemia determines the effect of type 2 diabetes risk genes on insulin secretion. *Diabetes* **2010**, *59*, 3247–3252. [CrossRef]

32. Herzberg-Schäfer, S.; Heni, M.; Stefan, N.; Häring, H.-U.; Fritsche, A. Impairment of GLP1-induced insulin secretion: Role of genetic background, insulin resistance and hyperglycaemia. *Diabetes Obes. Metab.* **2012**, *14* (Suppl. 3), 85–90. [CrossRef]

33. Gorgojo-Martínez, J.J. New glucose-lowering drugs for reducing cardiovascular risk in patients with type2 diabetes mellitus. *Hipertens. Riesgo Vasc.* **2019**, *36*, 145–161. [CrossRef] [PubMed]

34. Takayanagi, R.; Uchida, T.; Kimura, K.; Yamada, Y. Evaluation of Drug Efficacy of GLP-1 Receptor Agonists and DPP-4 Inhibitors Based on Target Molecular Binding Occupancy. *Biol. Pharm. Bull.* **2018**, *41*, 153–157. [CrossRef]

35. Lee, S.; Lee, D.Y. Glucagon-like peptide-1 and glucagon-like peptide-1 receptor agonists in the treatment of type 2 diabetes. *Ann. Pediatric Endocrinol. Metab.* **2017**, *22*, 15–26. [CrossRef] [PubMed]

36. Marbury, T.C.; Flint, A.; Jacobsen, J.B.; Derving Karsbøl, J.; Lasseter, K. Pharmacokinetics and Tolerability of a Single Dose of Semaglutide, a Human Glucagon-Like Peptide-1 Analog, in Subjects with and without Renal Impairment. *Clin. Pharmacokinet.* **2017**, *56*, 1381–1390. [CrossRef]

37. Gerstein, H.C.; Colhoun, H.M.; Dagenais, G.R.; Diaz, R.; Lakshmanan, M.; Pais, P.; Probstfield, J.; Riesmeyer, J.S.; Riddle, M.C.; Rydén, L.; et al. Dulaglutide and cardiovascular outcomes in type 2 diabetes (REWIND): A double-blind, randomised placebo-controlled trial. *Lancet* **2019**, *394*, 121–130. [CrossRef]

38. Nauck, M.A.; Meier, J.J.; Cavender, M.A.; el Aziz, M.A.; Drucker, D.J. Cardiovascular actions and clinical outcomes with glucagon-like peptide-1 receptor agonists and dipeptidyl peptidase-4 inhibitors. *Circulation* **2017**, *136*, 849–870. [CrossRef]

39. Thomas, M.C. The potential and pitfalls of GLP-1 receptor agonists for renal protection in type 2 diabetes. *Diabetes Metab.* **2017**, *43* (Suppl. 1), S20–S27. [CrossRef]

40. Greco, E.V.; Russo, G.; Giandalia, A.; Viazzi, F.; Pontremoli, R.; de Cosmo, S. GLP-1 Receptor Agonists and Kidney Protection. *Medicina* **2019**, *55*, 233. [CrossRef] [PubMed]

41. Brown, A.; Guess, N.; Dornhorst, A.; Taheri, S.; Frost, G. Insulin-associated weight gain in obese type 2 diabetes mellitus patients: What can be done? *Diabetes Obes. Metab.* **2017**, *19*, 1655–1668. [CrossRef]

42. Ramirez, G.; Morrison, A.; Bittle, P. Clinical practice considerations and review of the literature for the use of DPP-4 inhibitors in patients with type 2 diabetes and chronic kidney disease. *Endocr. Pract.* **2013**, *19*, 1025–1034. [CrossRef] [PubMed]

43. Scheen, A.J. Cardiovascular effects of new oral glucose-lowering agents DPP-4 and SGLT-2 inhibitors. *Circ. Res.* **2018**, *122*, 1439–1459. [CrossRef] [PubMed]

44. Fernandez-Fernandez, B.; Fernandez-Prado, R.; Górriz, J.L.; Martinez-Castelao, A.; Navarro-González, J.F.; Porrini, E.; Soler, M.J.; Ortiz, A. Canagliflozin and Renal Events in Diabetes with Established Nephropathy Clinical Evaluation and Study of Diabetic Nephropathy with Atrasentan: What was learned about the treatment of diabetic kidney disease with canagliflozin and atrasentan? *Clin. Kidney J.* **2019**, *12*, 313–321. [CrossRef] [PubMed]

45. Sarafidis, P.; Ferro, C.J.; Morales, E.; Ortiz, A.; Malyszko, J.; Hojs, R.; Khazim, K.; Ekart, R.; Valdivielso, J.; Fouque, D.; et al. SGLT-2 inhibitors and GLP-1 receptor agonists for nephroprotection and cardioprotection in patients with diabetes mellitus and chronic kidney disease. A consensus statement by the EURECA-m and the DIABESITY working groups of the ERA-EDTA. *Nephrol. Dial. Transplant.* **2019**, *34*, 208–230. [CrossRef] [PubMed]

46. Kapitza, C.; Nosek, L.; Jensen, L.; Hartvig, H.; Jensen, C.B.; Flint, A. Semaglutide, a once-weekly human GLP-1 analog, does not reduce the bioavailability of the combined oral contraceptive, ethinylestradiol/levonorgestrel. *J. Clin. Pharmacol.* **2015**, *55*, 497–504. [CrossRef] [PubMed]

47. Mosenzon, O.; Blicher, T.M.; Rosenlund, S.; Eriksson, J.W.; Heller, S.; Hels, O.H.; Pratley, R.; Sathyapalan, T.; Desouza, C.; PIONEER 5 Investigators. Efficacy and safety of oral semaglutide in patients with type 2 diabetes and moderate renal impairment (PIONEER 5): A placebo-controlled, randomised, phase 3a trial. *Lancet Diabetes Endocrinol.* **2019**, *7*, 515–527. [CrossRef]

48. Idorn, T.; Knop, F.K.; Jørgensen, M.B.; Jensen, T.; Resuli, M.; Hansen, P.M.; Christensen, K.B.; Holst, J.J.; Hornum, M.; Feldt-Rasmussen, B. Safety and Efficacy of Liraglutide in Patients with Type 2 Diabetes and End-Stage Renal Disease: An Investigator-Initiated, Placebo-Controlled, Double-Blind, Parallel-Group, Randomized Trial. *Diabetes Care* **2016**, *39*, 206–213. [CrossRef] [PubMed]

49. Davies, M.J.; Bain, S.C.; Atkin, S.L.; Rossing, P.; Scott, D.; Shamkhalova, M.S.; Bosch-Traberg, H.; Syrén, A.; Umpierrez, G.E. Efficacy and Safety of Liraglutide Versus Placebo as Add-on to Glucose-Lowering Therapy in Patients With Type 2 Diabetes and Moderate Renal Impairment (LIRA-RENAL): A Randomized Clinical Trial. *Diabetes Care* **2015**, *39*, 222–230. [CrossRef]

50. Pfeffer, M.A.; Claggett, B.; Diaz, R.; Dickstein, K.; Gerstein, H.C.; Køber, L.V.; Lawson, F.C.; Ping, L.; Wei, X.; Lewis, E.F.; et al. Lixisenatide in patients with type 2 diabetes and acute coronary syndrome. *N. Engl. J. Med.* **2015**, *373*, 2247–2257. [CrossRef] [PubMed]

51. Holman, R.R.; Bethel, M.A.; Mentz, R.J.; Thompson, V.P.; Lokhnygina, Y.; Buse, J.B.; Chan, J.C.N.; Choi, J.; Gustavson, S.M.; Iqbal, N.; et al. Effects of Once-Weekly Exenatide on Cardiovascular Outcomes in Type 2 Diabetes. *N. Engl. J. Med.* **2017**, *377*, 1228–1239. [CrossRef] [PubMed]

52. Marso, S.P.; Daniels, G.H.; Brown-Frandsen, K.; Kristensen, P.; Mann, J.F.; Nauck, M.A.; Nissen, S.E.; Pocock, S.; Poulter, N.R.; Ravn, L.S.; et al. Liraglutide and Cardiovascular Outcomes in Type 2 Diabetes. *N. Engl. J. Med.* **2016**, *375*, 311–322. [CrossRef] [PubMed]

53. Hernandez, A.F.; Green, J.B.; Janmohamed, S.; D'Agostino, R.B.; Granger, C.B.; Jones, N.P.; Leiter, L.A.; E Rosenberg, A.; Sigmon, K.N.; Somerville, M.C.; et al. Albiglutide and cardiovascular outcomes in patients with type 2 diabetes and cardiovascular disease (Harmony Outcomes): A double-blind, randomised placebo-controlled trial. *Lancet* **2018**, *392*, 1519–1529. [CrossRef]

54. Husain, M.; Birkenfeld, A.L.; Donsmark, M.; Dungan, K.; Eliaschewitz, F.G.; Franco, D.R.; Jeppesen, O.K.; Lingvay, I.; Mosenzon, O.; Pedersen, S.D.; et al. Oral Semaglutide and Cardiovascular Outcomes in Patients with Type 2 Diabetes. *N. Engl. J. Med.* **2019**, *381*, 841–851. [CrossRef] [PubMed]

55. Cefalu, W.T.; Kaul, S.; Gerstein, H.; Holman, R.R.; Zinman, B.; Skyler, J.S.; Green, J.B.; Buse, J.B.; Inzucchi, S.E.; Leiter, L.A.; et al. Cardiovascular Outcomes Trials in Type 2 Diabetes: Where Do We Go From Here? Reflections From aDiabetes CareEditors' Expert Forum. *Diabetes Care* **2017**, *41*, 14–31. [CrossRef] [PubMed]

56. Kristensen, S.L.; Rørth, R.; Jhund, P.S.; Docherty, K.F.; Sattar, N.; Preiss, D.; Køber, L.; Petrie, M.C.; McMurray, J.J.V. Cardiovascular, mortality, and kidney outcomes with GLP-1 receptor agonists in patients with type 2 diabetes: A systematic review and meta-analysis of cardiovascular outcome trials. *Lancet Diabetes Endocrinol.* **2019**, *7*, 776–785. [CrossRef]

57. Muskiet, M.; Tonneijck, L.; Huang, Y.; Liu, M.; Saremi, A.; Heerspink, H.J.L.; Van Raalte, D.H. Lixisenatide and renal outcomes in patients with type 2 diabetes and acute coronary syndrome: An exploratory analysis of the ELIXA randomised, placebo-controlled trial. *Lancet Diabetes Endocrinol.* **2018**, *6*, 859–869. [CrossRef]

58. Bethel, M.A.; Mentz, R.J.; Merrill, P.; Buse, J.B.; Chan, J.C.N.; Goodman, S.G.; Iqbal, N.; Jakuboniene, N.; Katona, B.G.; Lokhnygina, Y.; et al. Renal Outcomes in the EXenatide Study of Cardiovascular Event Lowering (EXSCEL). *Diabetes* **2018**, *67*, 522. [CrossRef]

59. Perkovic, V.; Koitka-Weber, A.; E Cooper, M.; Schernthaner, G.; Pfarr, E.; Woerle, H.J.; Von Eynatten, M.; Wanner, C. Choice of endpoint in kidney outcome trials: Considerations from the EMPA-REG OUTCOME®trial. *Nephrol. Dial. Transplant.* **2019**. [CrossRef]

60. Perkovic, V.; Bain, S.; Bakris, G.; Buse, J.; Gondolf, T. eGFR loss with glucagon-like peptide-1 (GLP-1) analogue treatment: Data from SUSTAIN 6 and LEADER. In Proceedings of the 56th ERAEDTA Congress, Budapest, Hungary, 13–16 June 2019; pp. 13–16.

61. Pyke, C.; Heller, R.S.; Kirk, R.K.; Ørskov, C.; Reedtz-Runge, S.; Kaastrup, P.; Hvelplund, A.; Bardram, L.; Calatayud, D.; Knudsen, L.B. GLP-1 Receptor Localization in Monkey and Human Tissue: Novel Distribution Revealed With Extensively Validated Monoclonal Antibody. *Endocrinol.* **2014**, *155*, 1280–1290. [CrossRef]

62. Pyke, C.; Knudsen, L.B. The glucagon-like peptide-1 receptor-or not? *Endocrinology* **2013**, *154*, 4–8. [CrossRef]

63. Tsimihodimos, V.; Elisaf, M. Effects of incretin-based therapies on renal function. *Eur. J. Pharmacol.* **2018**, *818*, 103–109. [CrossRef] [PubMed]

64. Katout, M.; Rutsky, J.; Shah, P.; Zhu, H.; Brook, R.D.; Zhong, J.; Rajagopalan, S. Effect of GLP-1 Mimetics on Blood Pressure and Relationship to Weight Loss and Glycemia Lowering: Results of a Systematic Meta-Analysis and Meta-Regression. *Am. J. Hypertens.* **2013**, *27*, 130–139. [CrossRef] [PubMed]

65. Wang, L.; Li, P.; Tang, Z.; Yan, X.; Feng, B. Structural modulation of the gut microbiota and the relationship with body weight: Compared evaluation of liraglutide and saxagliptin treatment. *Sci. Rep.* **2016**, *6*, 33251. [CrossRef] [PubMed]

66. Gutzwiller, J.-P.; Tschopp, S.; Bock, A.; Zehnder, C.E.; Huber, A.R.; Kreyenbuehl, M.; Gutmann, H.; Drewe, J.; Henzen, C.; Goeke, B.; et al. Glucagon-Like Peptide 1 Induces Natriuresis in Healthy Subjects and in Insulin-Resistant Obese Men. *J. Clin. Endocrinol. Metab.* **2004**, *89*, 3055–3061. [CrossRef] [PubMed]

67. Sloan, L.A. Review of glucagon-like peptide-1 receptor agonists for the treatment of type 2 diabetes mellitus in patients with chronic kidney disease and their renal effects. *J. Diabetes* **2019**, *11*, 938–948. [CrossRef] [PubMed]

68. Thomson, S.C.; Kashkouli, A.; Singh, P. Glucagon-like peptide-1 receptor stimulation increases GFR and suppresses proximal reabsorption in the rat. *Am. J. Physiol. Ren. Physiol.* **2013**, *304*, F137–F144. [CrossRef]

69. Tonneijck, L.; Smits, M.M.; Muskiet, M.; Hoekstra, T.; Kramer, M.H.H.; Danser, A.H.J.; Diamant, M.; Joles, J.A.; Van Raalte, D.H. Acute renal effects of the GLP-1 receptor agonist exenatide in overweight type 2 diabetes patients: A randomised, double-blind, placebo-controlled trial. *Diabetol.* **2016**, *59*, 1412–1421. [CrossRef]

70. Yin, W.-Q.; Xu, S.; Wang, Z.; Liu, H.; Peng, L.; Fang, Q.; Deng, T.; Zhang, W.; Lou, J. Recombinant human GLP-1(rhGLP-1) alleviating renal tubulointestitial injury in diabetic STZ-induced rats. *Biochem. Biophys. Res. Commun.* **2018**, *495*, 793–800. [CrossRef]
71. Study, F. Study to See How Semaglutide Works Compared to Placebo in People with Type 2 Diabetes and Chronic Kidney Disease (FLOW). ClinicalTrials.gov Identifier: NCT0383. 1915. Available online: https://clinicaltrials.gov/ct2/show/NCT0383 (accessed on 28 March 2020).

Review

Calciprotein Particles and Serum Calcification Propensity: Hallmarks of Vascular Calcifications in Patients with Chronic Kidney Disease

Ciprian N. Silaghi [1,*], Tamás Ilyés [1], Adriana J. Van Ballegooijen [2] and Alexandra M. Crăciun [1]

[1] Department of Molecular Sciences, University of Medicine and Pharmacy "Iuliu Hațieganu", 400012 Cluj-Napoca, Romania; tamasilyes94@gmail.com (T.I.); acraciun@umfcluj.ro (A.M.C.)

[2] Department of Nephrology & Epidemiology and Biostatistics, Amsterdam UMC, location VUmc, 1117 HV Amsterdam, Netherlands; aj.vanballegooijen@vumc.nl

* Correspondence: silaghi.ciprian@umfcluj.ro

Received: 1 April 2020; Accepted: 29 April 2020; Published: 29 April 2020

Abstract: Cardiovascular complications are one of the leading causes of mortality worldwide and are strongly associated with atherosclerosis and vascular calcification (VC). Patients with chronic kidney disease (CKD) have a higher prevalence of VC as renal function declines, which will result in increased mortality. Serum calciprotein particles (CPPs) are colloidal nanoparticles that have a prominent role in the initiation and progression of VC. The T_{50} test is a novel test that measures the conversion of primary to secondary calciprotein particles indicating the tendency of serum to calcify. Therefore, we accomplished a comprehensive review as the first integrated approach to clarify fundamental aspects that influence serum CPP levels and T_{50}, and to explore the effects of CPP and calcification propensity on various chronic disease outcomes. In addition, new topics were raised regarding possible clinical uses of T_{50} in the assessment of VC, particularly in patients with CKD, including possible opportunities in VC management. The relationships between serum calcification propensity and cardiovascular and all-cause mortality were also addressed. The review is the outcome of a comprehensive search on available literature and could open new directions to control VC.

Keywords: calciprotein particles; calcification propensity; chronic kidney disease; vascular calcification

---

1. Introduction

Serum calciprotein particles (CPPs) are colloidal nanoparticles comprising a combination of proteins (mainly fetuin-A, but also albumin and Gla-rich protein (GRP)) and calcium (Ca^{2+}) containing compounds, primarily calcium phosphate [1–3]. They are first formed by the binding of Ca^{2+} precursors to the acidic residues of fetuin-A, a glycoprotein secreted by the liver [1,4]. These calcium–protein complexes, also known as calciprotein monomers, pass through further aggregation and maturation, resulting in primary calciprotein particles (CPP I) and later on, secondary calciprotein particles (CPP II) [5–7]. CPP I are small spherical colloidal nanoparticles that contain amorphous calcium phosphate, while CPP II contain crystalline calcium phosphate at their core, are larger than CPP I, and have a needle-shaped structure. This transition from CPP I to CPP II is called "ripening" and is hypothesized to be attributed to a reorganization of the colloidal nanoparticles into a more stable form [5]. The ripening process is influenced by a number of factors such as the concentration of fetuin-A, Ca^{2+}, magnesium (Mg^{2+}), phosphate (Pi), as well as the temperature and pH of the surrounding microenvironment [1,6,8].

The transition from CPP I to CPP II, which takes place naturally in serum, can also be induced in vitro, and the time needed for the transition to take place can be measured. Half of the time needed for the spontaneous transition from CPP I to CPP II, designated as T_{50}, has been established as a strong predictor of the calcifying properties of serum [9]. A higher T_{50} is beneficial since serum with a higher T_{50} is less prone to calcify tissues compared to serum that has a lower T_{50}.

Vascular calcification (VC) results in the thickening and increased rigidity of muscular arterial walls [10]. This is the consequence of two main types of calcification: intimal and medial calcification. Intimal calcification is associated with atherosclerosis, Ca^{2+} being deposited along with lipoproteins as well as phospholipids [11,12]. Medial calcification, which is more prevalent in chronic kidney disease (CKD), is the result of an osteogenic process similar to intramembranous ossification, which is independent of atherosclerosis and causes a decrease in compliance of the vessel wall [13–15]. Medial calcification occurs earlier in CKD patients compared to the general population [16].

With respect to CPPs in general and T_{50} in particular, there have been no reviews published until now that summarize findings related to both CPPs and T_{50}. Therefore, the purpose of this review was to offer a synopsis of all studies published on CPPs and T_{50}, respectively. We also aim to analyse and discuss their roles and clinical significance in patients prone to developing VC, as well as to establish possible new directions in the management of VC.

2. Methodology

2.1. Search Strategy

All databases that could be accessed through the PubMed search engine were selected for this review. Human, animal, and in vitro studies were all taken into account. Due to the specific nature of the selected domain and the fact that the majority of research papers were published relatively recently, the period of publication was not limited. A set of search terms was selected as follows: "Calciprotein particles", "T_{50} AND calcification", "Serum calcification propensity". The search was performed in PubMed on the 4th of January 2020 for both search strings, yielding a total of 162 studies (78, 30, and 54 results, respectively). The results of the searches were organized into lists that were cross-checked between search terms, with duplicates being eliminated. After the initial screening of titles and abstracts, full-text articles were obtained for all eligible studies.

2.2. Selection, Screening, and Inclusion

The authors jointly selected the inclusion and exclusion criteria. Only articles with abstracts were selected for screening, written in English including human, animal, and in vitro studies.

Studies that did not address CPPs and/or T_{50} in a medically relevant manner, such as physical or chemical characterization of CPPs, and studies that lacked a clear definition of methods and materials were not included. Reviews and case reports were excluded as well.

The identification, selection, screening, and inclusion process is summarized in Figure 1. After cross-checking and eliminating duplicates, the results of the search string "Serum calcification propensity" yielded three studies that were subsequently included in the same category as T_{50}. In total, 18 studies were included for CPPs [3,17–33] and 30, including the aforementioned 3 studies, for T_{50} [34–63].

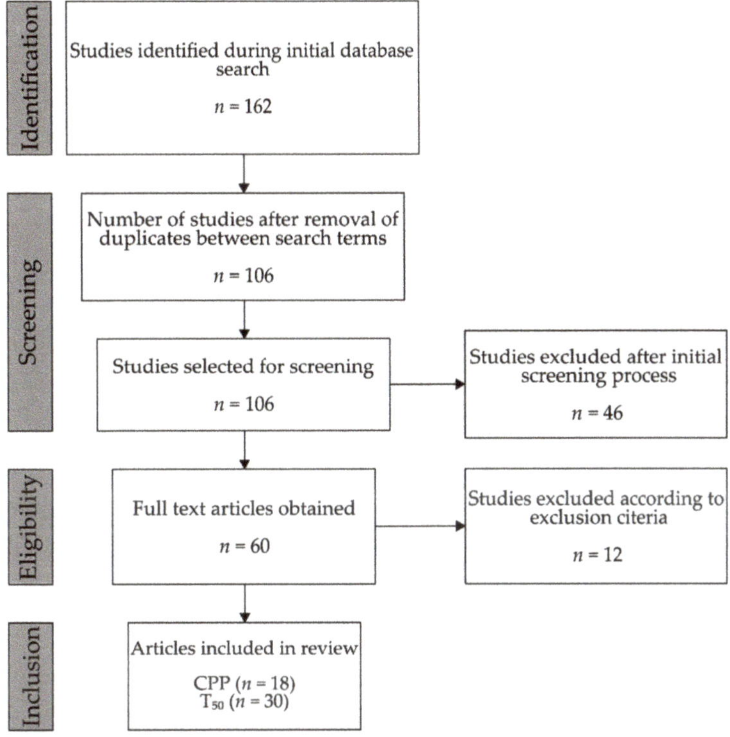

Figure 1. Flow diagram for the identification, selection, screening, and inclusion process. Abbreviations: CPP, calciprotein particles.

3. Molecular Background

3.1. Fetuin-A and Calciprotein Particles

While CPPs contain a number of proteins that can bind Ca^{2+}, e.g., Gla-rich protein (GRP) [3], as well as other serum proteins and lipoproteins such as albumin and apolipoprotein A1 [2], the main protein within the CPP structure is fetuin-A, also known as alpha-2-HS-glycoprotein. It is a 55–60 kDa glycoprotein, synthetized and secreted by the liver, which undergoes post-translational modifications, including phosphorylation [4,64]. While phosphorylation is crucial for its various interactions, e.g., with the insulin receptor, it is not required for mineral binding due to the number of acidic residues [1,4,65,66]. Each molecule of fetuin-A can bind up to 6 Ca^{2+} ions [67]. Calcium and Pi bound by fetuin-A form protein–mineral complexes called calciprotein monomers, the aggregation of which results in the formation of plasma-soluble amorphous colloidal particles, referred to as CPP I. The CPP I, which is spherical in nature and has a diameter of around 75 nm, circulates in plasma and eventually undergoes rearrangement into CPP II, which is more dense, with a larger diameter (120 nm), insoluble in serum, and has a needle-shaped crystalline structure [1]. This transition from the primary, more instable form, to the secondary, more stable form, is dubbed "ripening" [5]. The process is illustrated in Figure 2.

CPP I and CPP II are cleared by macrophages, especially Kupffer cells in the liver, thereby preventing tissular deposition of Ca^{2+} and Pi [68]. Studies have shown that CPP II induces vascular smooth muscle cell (VSMC) calcification in vitro, as well as the secretion of tumour necrosis factor α (TNF-α) in macrophages, while CPP I does not. CPP II was found to increase bone morphogenetic protein-2 as well as nuclear factor kappa-B expression in VSMCs. The calcification of VSMCs was also shown to be the result of the cellular uptake of CPP II, with CPP II being detected intracellularly in calcified VSMCs [29]. Both CPP I and CPP II were found to induce VSMC intimal hyperplasia, which was more pronounced in the case of CPP II [18]. Moreover, CPPs were found to induce secretion of interleukin 1β (IL-1β) in macrophages, however, to a lesser degree than hydroxyapatite crystals [31]. While both forms of CPP have pro-inflammatory effects, it is still less prominent than crystalline hydroxyapatite. The more pronounced pro-inflammatory effect of CPP II compared to that of CPP I might be attributed to its content of hydroxyapatite in crystalline form.

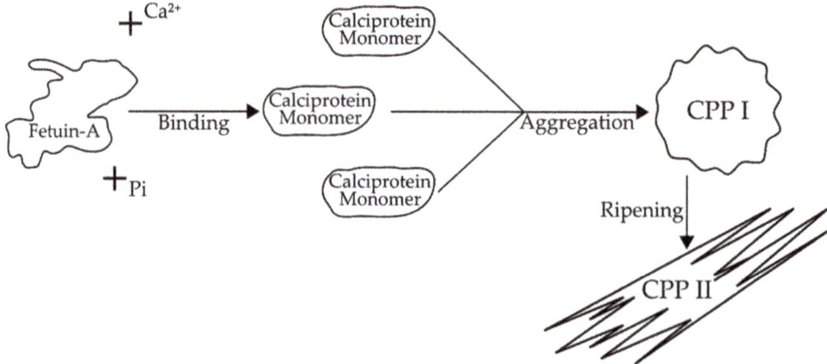

Figure 2. Fetuin-A transformation into CPP II. Abbreviations: Pi, phosphate; CPP I, primary calciprotein particle; CPP II, secondary calciprotein particle.

The CPPs are detected and quantified in serum indirectly, by assessing the fetuin-A levels via enzyme-linked immunosorbent assay (ELISA), before and after a high-speed centrifugation that precipitates all CPPs as CPP II. The difference between fetuin-A concentrations before and after centrifugation is interpreted as the amount of CPPs in the serum sample [33,69]. Because this method induces the ripening process before measuring CPP content, it only brings information regarding the total concentration of CPPs, without differentiating between CPP I and CPP II. To measure CPP I and CPP II concentrations independently, a flow-cytometry method can be used [70].

3.2. Calcifying Properties of Serum

A method for measuring the calcification inhibition capacity of serum was elaborated by Ismail et al. [71] based on electrochemical impedance. A prototype probe was successfully used to measure the impedance of a test solution consisting of bovine albumin, Ca^{2+}, and Pi. Upon the addition of a calcification inhibitor, in that case fetuin-A, the electrical impedance of the solution would increase proportionately to the Ca^{2+} content, due to the inhibitor consuming Ca^{2+} ions by forming CPP I. Thus, the calcification inhibition capacity of the serum could be determined by measuring the variation of impedance of a solution containing Ca^{2+} and Pi in a known concentration, after the addition of serum.

Pasch et al. [9] were the first to develop a plate-based nephelometric assay to measure the time needed for the transition from CPP I to CPP II in serum treated with Ca^{2+} and Pi solutions, and proposed the use of one half of the transition time to maximum turbidity, also known as T_{50}, as a parameter to describe the calcifying properties of serum. The influence of factors such as pH and concentrations of various serum constituents upon T_{50} was also analysed, and is summarized in Figure 3.

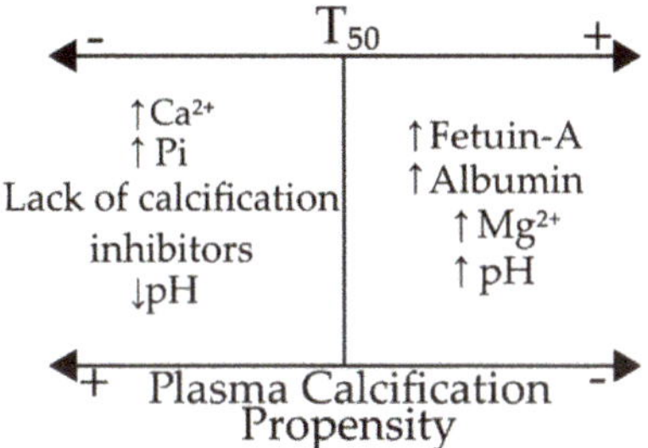

Figure 3. Effects of various factors upon T$_{50}$ (half of the time needed for the spontaneous transition from CPP I to CPP II) and plasma calcification propensity. Abbreviations: Pi, phosphate; Ca²⁺, calcium; Mg²⁺, magnesium.

4. Results

4.1. Calciprotein Particles

Human studies on serum CPP levels are summarized in Table 1, animal and in vitro studies on CPP are summarized in Table 2. The majority of studies used detection methods that did not differentiate between the two types of CPP. To avoid confusion, we used the term total CPP (tCPP) when referring to studies that did not specify the type of CPP analysed.

Table 1. Summary of 11 human studies on calciprotein particle (CPP).

Author, Year	Study Design, Duration	Number of Subjects, Disease	CPP Type Studied	Findings
Nakazato et al. 2019 [20]	cross-sectional, N/A	71 ACS	tCPP	High CPP levels associated with atherosclerosis.
Chen et al. 2019 [23]	cross-sectional, N/A	45 CKD stage IV–V	CPP II	Larger CPP II diameter in patients with VC.
Viegas et al. 2018 [3]	cross-sectional, N/A	16 CKD stage II-IV, 20 CKD stage V	tCPP	CPP from CKD stage V patients contained less fetuin-A and GRP and had CPP II like characteristics.
Yamada et al. 2018 [28]	cross-sectional, N/A	10 diabetes mellitus type 2	tCPP	CPP elevated 2 h post-meal, CPP inversely correlated with eGFR.
Cai et al. 2015 [30]	cross-sectional, N/A	20 peritoneal dialysis	tCPP	CPP present, fetuin-A abundant in peritoneal dialysis effluent.
Smith et al. 2013 [32]	cross-sectional, N/A	11 CKD stage III–IV, 42 HD, 18 peritoneal dialysis, 13 chronic inflammatory disease	tCPP	CPP increased in CKD III-IV, HD, peritoneal dialysis and chronic inflammatory disease patients; CPP was highest in HD patients with calcific uremic arteriolopathy.
Smith et al. 2012 [33]	cross-sectional, N/A	200 CKD stage III–IV	tCPP	Higher CPP levels associated with increased aortic stiffness.
Cai et al. 2018 [24]	prospective cohort, 7 weeks	12 peritoneal dialysis	tCPP	Dialysate with higher Ca²⁺ concentration had higher CPP content.
Ruderman et al. 2018 [25]	prospective cohort, 12 months	62 HD	CPP I	Increase of serum CPP I after cessation of cinacalcet treatment.
Bressendorff et al. 2019 [17]	Interventional, 28 days	57 HD	CPP I, CPP II	Higher Mg²⁺ concentration dialysis solution reduced both CPP I and CPP II levels, compared to standard dialysis solution.
Nakamura et al. 2019 [21]	Interventional, 16 weeks	24 HD	tCPP	Lower CPP in lanthanum carbonate treated patients vs. calcium carbonate.

Abbreviations: HD, haemodialysis; ACS, acute coronary syndrome; CPP, calciprotein particle; CPP I, primary calciprotein particle; CPP II, secondary calciprotein particle; tCPP, total calciprotein particles; CKD, chronic kidney disease; VC, vascular calcification; GRP, Gla-rich protein; eGFR, estimated glomerular filtration rate; N/A, not applicable.

Table 2. Summary of 1 animal and 6 in vitro studies on CPP.

Author, Year	Study Design	Animals/Cells	CPP Type Studied	Findings
Nemoto et al. 2019 [22]	animal	rats with 5/6 nephrectomy	tCPP	Lower CPP in rats treated with sucroferric oxyhydroxide.
Shishkova et al. 2019 [18]	in vitro	VSMCs	CPP I, CPP II	Both CPP I and CPP II induced VSMC intimal hyperplasia, more pronounced in case of CPP II.
Ter Braake et al. 2019 [19]	in vitro	VSMCs	CPP II	CPP II induced VSMC calcification.
Aghagolzadeh et al. 2017 [26]	in vitro	VSMCs	tCPP	H_2S inhibits CPP induced VSMC calcification.
Cai et al. 2017 [27]	in vitro	VSMCs	CPP II	Pi or CPP II alone did not initiate VSMC mineralization, but CPP II with Pi did.
Aghagolzadeh et al. 2016 [29]	in vitro	VSMCs	CPP I, CPP II	CPP II induced calcification in VSMCs, CPP I did not.
Smith et al. 2013 [31]	in vitro	VSMCs	tCPP	CPP induce secretion of TNF-α and IL-1β in macrophages, but less significantly than that induced by hydroxyapatite crystals.

Abbreviations: VSMCs, vascular smooth muscle cells; CPP, calciprotein particle; CPP I, primary calciprotein particle; CPP II, secondary calciprotein particle; tCPP, total calciprotein particles; H_2S, hydrogen sulphide; Pi, phosphate; TNF-α, tumour necrosis factor α; IL-1β, interleukin 1β.

Dialysate from haemodialysis (HD) patients was found to contain CPP, and higher dialysate Ca^{2+} content was found to be associated with higher CPP concentration [24,30]. This suggests that CPP can be cleared from the plasma of patients with chronic kidney disease (CKD) through HD. In addition, CPP were found to induce VSMC calcification and intimal hyperplasia, with higher serum levels of CPP being associated with increased aortic stiffness [18,26,33]. CPP also induced the secretion of TNF-α and IL-1β in macrophages, with a more pronounced effect being attributed to CPP II. This pro-inflammatory response, however, was still inferior to that induced by pure hydroxyapatite crystals [29,31].

4.2. Calcification Propensity

Observational studies of T_{50} and outcomes are summarized in Table 3, and human intervention studies are summarized in Table 4. The majority of studies included in this section concern T_{50} in CKD and/or kidney transplant patients.

Oral Mg^{2+} supplementation, as well as increased Mg^{2+} concentration in dialysis solution was found to increase T_{50} in CKD patients [39,45,51]. The T_{50} was also found to be associated with serum Mg^{2+} levels in CKD patients, but not with eGFR [50]. Serum Mg^{2+} levels were directly associated with T_{50}, which suggests that both oral Mg^{2+} supplementation, as well as increasing the Mg^{2+} content of dialysis solution could be a viable method to counterbalance VC to some extent in CKD patients. The use of citrate-buffered dialysis solution was found to significantly increase T_{50} as opposed to standard acetate-buffered dialysis solution in HD patients [34,46]. While platelet derived growth factor B hypomorphic animal brains showed signs of calcification, T_{50} did not differ compared to controls [61].

Lower T_{50} levels were also found to be associated with lower tissue oxygenation, as well as an increase in all-cause and cardiovascular mortality, especially in CKD and kidney transplant patients [36,42,48,49,52,56,57,60].

Table 3. Summary of 18 observational studies on T_{50} and health outcomes.

Author, Year	Study Design	Follow-Up Time	Number of Subjects, Disease	Findings
Bullen et al. 2019 [41]	cross-sectional	N/A	149 men with osteoporosis	T_{50} was not associated with bone mineral density.
Dahdal et al. 2018 [47]	cross-sectional	N/A	168, SLE	T_{50} was negatively associated with disease activity.
Pruijm et al. 2017 [48]	cross-sectional	N/A	58, CKD; 48, hypertension	Lower T_{50} was associated with reduced tissue oxygenation and perfusion.
Bielesz et al. 2017 [50]	cross-sectional	N/A	118, CKD stage I–V	T_{50} associated with Pi, Mg^{2+} and fetuin-A but not with eGFR.
Dekker et al. 2016 [54]	cross-sectional	N/A	64, HD	T_{50} increased post-haemodialysis and post-haemodiafiltration.
Voelkl et al. 2018 [63]	cross-sectional	N/A	16, CKD; 20, HD	T_{50} was lower in CKD patients compared to controls.
van Dijk et al. 2019 [35]	prospective cohort	15 years	216, type 1 diabetes	T_{50} not associated with mortality.
Bundy et al. 2019 [36]	prospective cohort	At TOD or 11.2 years	3404, CKD stage II–IV	Lower T_{50} associated with cardiovascular events and all-cause mortality.
Ponte et al. 2019 [37]	prospective cohort	3 months	46, HD; 12, peritoneal dialysis	Higher T_{50} after dialysis initiation.
Bundy et al. 2019 [38]	prospective cohort	3.2 ± 0.6 years	780, CKD stage II–IV	Lower T_{50} was associated with greater CAC severity and progression, however, T_{50} was not associated with CAC incidence.
Bostom et al. 2018 [42]	prospective cohort	median of 2.18 years	685, CVD	Lower T_{50} and fetuin-A levels were associated with greater risk for CVD outcomes.
Pasch et al. 2017 [49]	prospective cohort	At TOD or first non-fatal CVE	2785, HD	Lower T_{50} associated with all-cause mortality, myocardial infarction, and peripheral vascular events.
Lorenz et al. 2017 [52]	prospective cohort	24 months	188, HD	T_{50} rate of decline significantly predicted all-cause and cardiovascular mortality.
Dahle et al. 2016 [56]	prospective cohort	median of 5.1 years	1435, kidney transplant	Lower T_{50} associated with all-cause and cardiac mortality.
Keyzer et al. 2016 [57]	prospective cohort	median of 3.1 years	699, kidney transplant	Lower T_{50} associated with increased graft failure, all-cause, and cardiac mortality.
de Seigneux et al. 2015 [59]	prospective cohort	1 year	21, kidney donors	T_{50} was independent of eGFR.
Smith et al. 2014 [60]	prospective cohort	median of 5.3 years	184, CKD stage III–IV	Lower T_{50} associated with higher all-cause mortality.
Berchtold et al. 2016 [58]	retrospective cohort	between 2 and 43 years	129, kidney transplant	T_{50} associated with interstitial fibrosis and vascular lesions.

Abbreviations: SLE, systemic lupus erythematosus; HD, haemodialysis; CKD, chronic kidney disease; CAC, coronary artery calcification; CVD, cardiovascular disease; Mg^{2+}, magnesium; TOD, time of death; CVE, cardiovascular event; Pi, phosphate; eGFR, estimated glomerular filtration rate; N/A, not applicable.

Table 4. Summary of 11 human interventional studies on T_{50} with health outcomes.

Author, Year	Study Duration	Number of Subjects, Disease	Findings
Smerud et al. 2017 [53]	1 year	123, kidney transplant	T_{50} increased with no further change after 1 year, ibandronate had no effect on T_{50}.
Andrews et al. 2018 [43]	12 weeks	80, CKD with hyperuricemia	Allopurinol lowered uric acid levels but had no effect on T_{50}.
Lorenz et al. 2018 [46]	3 months	78, HD	Acetate-free, citrate-acidified, standard bicarbonate dialysis solution increased T_{50} compared to acetate dialysis solution.
Ussif et al. 2018 [44]	1 year	76, kidney transplant	Paricalcitol supplementation had no effect on T_{50}.
Bressendorff et al. 2018 [45]	28 days	57, HD	Higher dialysis solution Mg^{2+} concentration increased T_{50}.
Bristow et al. 2016 [55]	3 months	41, post-menopausal women	Insignificant decrease of T_{50} in the group treated with oral calcium carbonate supplement.
Bressendorff et al. 2017 [51]	8 weeks	36, CKD III–IV	Oral Mg^{2+} supplementation increased T_{50}.
Aigner et al. 2019 [40]	4 weeks	35, CKD	Oral bicarbonate supplementation showed no effect on T_{50} in acidotic CKD patients.
Kendrick et al. 2018 [62]	14 weeks	18, CKD	Oral sodium bicarbonate supplementation showed no effect on T_{50} in CKD patients with low serum bicarbonate levels.
Ter Meulen et al. 2019 [34]	2 weeks	18, HD	Citric acid-buffered dialysis solution increased T_{50} compared to acetate-buffered solution.
Quiñones et al. 2019 [39]	2 weeks	9, CKD stage III, 9, CKD stage V	Effervescent, oral, calcium-magnesium citrate increased T_{50}.

Abbreviations: HD, haemodialysis; CKD, chronic kidney disease; Mg^{2+}, magnesium.

5. Discussion

This comprehensive review showed that multiple lines of evidence (cell, animal, and human) indicate that T_{50} is shorter in CKD and dialysis populations. A large amount of studies indicate that a lower T_{50} is related to VC, cardiovascular events, and mortality. These findings are robust across various populations and open up new directions to modify VC especially in patients with CKD. One of these factors that can influence the tendency to calcify is Mg^{2+}. Oral Mg^{2+} supplementation as well as increased dialysis solution Mg^{2+} concentration had beneficial effects on T_{50} [39,45,52], and a lower T_{50} was associated with cardiovascular and all-cause mortality in various populations [36,42,49,52,56,57,60]. It is worth noting the correlation between higher serum CPP content, especially CPP II, and VSMC inflammation as well as calcification [18,26,29,31,33]. Taking the included studies into consideration, we address two topics for further research in this relatively recent domain.

5.1. The Effect of Dialysis Solution Composition upon Serum Calcification Propensity in CKD Patients

The transition from CPP I to CPP II is delayed by the presence of Mg^{2+}, this effect being dependent upon the concentration of Mg^{2+}. The presence of Mg^{2+}, however, does not inhibit VSMC calcification in the presence of CPP II, suggesting that the anti-calcific effects of Mg^{2+} are more related to preventing the transition from CPP I to CPP II [19]. This would also explain the effect of Mg^{2+} upon increasing T_{50}. However, the exact mechanism by which Mg^{2+} inhibits the maturation of CPP I is not completely understood. One possible mechanism might lie in the ability of Mg^{2+} to inhibit Ca^{2+} and Pi crystallization [72], which is a necessary step in CPP maturation.

Studies suggest that there is a significant amount of CPPs in the dialysate of CKD patients on peritoneal dialysis. That CPP content was also directly proportional to the dialysate's Ca^{2+} content [24]. While HD was found to increase T_{50}, thus reducing the calcification propensity of the patient's plasma [37,54], serum CPP I and CPP II levels seem to be unaffected by standard HD [17].

First of all, this would suggest that the increase in T_{50} after initiation of HD is not attributed to the clearance of CPPs per se, but to the reduction of factors that precipitate the ripening process, most probably the reduction of Ca^{2+} and Pi. Second of all, CPPs, while not being cleared from the serum under standard HD conditions, are cleared by peritoneal dialysis to some degree. However, if the Mg^{2+} concentration of HD dialysis solution is increased, CPPs appear to pass the dialysis membrane and are cleared from the patient's serum [17]. This would, in part, explain the significant increase of T_{50} in patients treated with a dialysis solution containing a larger Mg^{2+} concentration compared to standard solution [45].

In addition to the beneficial effect of increased Mg^{2+} content in dialysis solution upon the serum calcification propensity in CKD patients, the use of an acetate-free, citrate-acidified dialysis solution was also found to increase T_{50} thus reducing the calcification propensity [34,46].

Patients with CKD who received oral Mg^{2+} supplementation showed a significant increase in T_{50} [39,51]. In post-menopausal women, the introduction of oral Ca^{2+} supplementation showed a decrease in T_{50}, however, this decrease did not differ significantly from the control group [55]. These observations correspond with the findings of Pasch et al. [9], who determined that higher serum Mg^{2+} levels will increase T_{50}. A summary of the aforementioned factors upon T_{50} is presented in Figure 4.

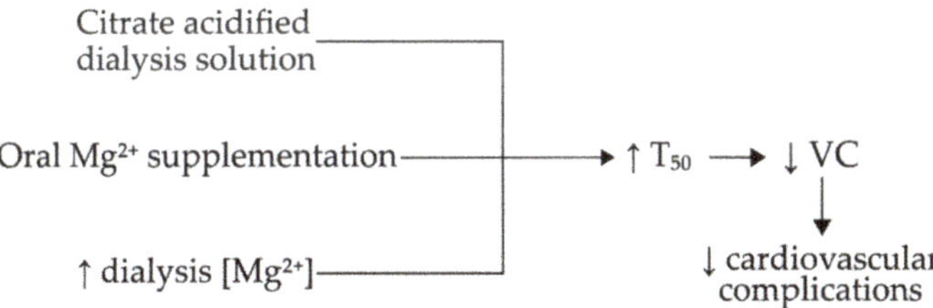

Figure 4. Summary of factors that increase T_{50} in CKD patients. Abbreviations: Mg, magnesium; VC, vascular calcification.

Furthermore, it is well known that patients with CKD have a significantly higher risk for VC and associated cardiovascular mortality [73]. Developing a standardized treatment plan for end-stage CKD patients on HD or peritoneal dialysis that would take into account the above outlined criteria, namely the increased Mg^{2+} content of dialysis solution and the use of citrate instead of acetate, should be validated and subsequently introduced into a therapeutic protocol. Patients with HD, as well as those with CKD who do not require HD, could also benefit from a reduction in oral Ca^{2+} and an increase in oral Mg^{2+} supplementation, respectively. Such an approach to the management of VC and the possible ensuing reduction of cardiovascular mortality rates in CKD patients could lead to an increased quality of life, especially for patients undergoing HD or peritoneal dialysis, delaying the onset or decreasing the severity of cardiovascular complications associated with CKD.

5.2. The T_{50} Test Could Be Used as a Factor in the Staging and/or Prognosis of CKD

There are plentiful studies, conducted on large sample sizes, that came to the conclusion that lower T_{50} corresponding to higher calcification propensity is strongly associated with increased cardiovascular and all-cause mortality rates [36,42,49,52,56,57,60]. Lower T_{50} was also associated with coronary artery calcification progression as well as greater risk for cardiovascular disease outcomes, such as myocardial infarction and peripheral vascular events [38,42,49].

The investigation of a possible association between T_{50} and eGFR could lead to the development of a reference interval for T_{50} in CKD patients, which is dependent on CKD stage. Such a reference interval, which has not yet been established, could be used as an additional prognostic parameter for CKD patients, especially those undergoing HD or peritoneal dialysis treatment. There was conflicting evidence that links serum CPP levels and T_{50} to eGFR. Yamada et al. [28] found that CPP levels were inversely associated with eGFR in diabetic patients. However, that study was conducted on diabetic patients, not CKD patients, and the patient group was relatively small as well. On the other hand, Bielesz et al. [50], found that T_{50} was not associated with eGFR in CKD stage I–V patients, instead being associated with numerous parameters, including Pi and Mg^{2+} levels. A similar result was obtained by de Seigneux et al. [59], who discovered that T_{50} was independent of eGFR in kidney transplant donors, which could be attributed to the compensation effect of an otherwise healthy remaining kidney. Those studies clearly pointed that while serum CPP levels are correlated with eGFR, T_{50} was not.

The CPP levels and T_{50} do not seem to be directly correlated with one another, although T_{50} is greatly influenced by serum Ca^{2+} levels and, in addition, CPP levels are directly proportional to circulating Ca^{2+} levels. Considering the previously discussed ideas, it could be hypothesized that CPP levels are correlated with T_{50}, justifying further studies in larger populations to investigate the association between T_{50} and eGFR. However, until the completion of this review, no studies have identified this relationship.

Even in the absence of a link between T_{50} and eGFR, but in the context of association between higher serum calcification propensity and increased cardiovascular and all-cause mortality rates especially in CKD patients, the use of T_{50} as risk factor that can be monitored should be considered. The ensuing introduction of measures to decrease calcification propensity could significantly reduce VC and related mortality in CKD patients.

An interesting opportunity would be to expand the area of research towards the involvement of CPPs in the calcification paradox, in which the presence of vascular calcification overlaps at the same time with bone demineralization assessed by a decrease in bone mineral density (BMD) [74]. It is difficult to decode how CPP and the interplay between vasculature–bone–kidney underlie the deleterious effect of calcification. On one hand, fetuin-A accumulates in calcified atherosclerotic plaques [75], but also in bone where it inhibits mineralization and halts bone matrix protein expression [76]. On the other hand, serum levels of fetuin-A were found to be decreased in patients with end-stage renal disease [77]. Contrariwise, serum CPP increases in patients with CKD III–IV, with it being the highest in HD patients [32] but with less fetuin-A content as CKD stage worsens [3]. Probably the turn-over of CPP is accelerated in CKD patients, but fetuin-A is consumed exerting its systemic anti-calcification effect necessary to counteract VC as CKD stage aggravates.

In addition to the well-known presence of VC in patients with CKD, an important decrease of BMD was also reported [78]. In maintenance HD patients, serum fetuin-A was inversely associated with coronary artery calcification and positively with BMD [79]. In respect to VC, serum CPP appears to behave divergently regarding fetuin-A dynamics in CKD patients: higher CPP levels are associated with increased aortic stiffness [33] and larger CPP II diameters were found in patients with VC [23]. As might be expected, T_{50} was inversely associated with coronary artery calcification (CAC) severity in CKD patients [38], thereby, the T_{50} test seems to mimic serum fetuin-A variations in respect to VC, as they were found to be associated [50]. Regarding the loss of skeletal mineral, T_{50} was not associated with BMD [38] and in the case of CPPs we did not find conclusive studies. To make the puzzle even more complicated, we could introduce the relationship between CPPs or T_{50} and eGFR, as discussed above. Thereby, CPPs were found to be inversely correlated with eGFR [28], instead of T_{50}, which was independent of eGFR [50,59].

However, an attempt to explain the paradox of calcification on the vasculature–bone–kidney axis only in terms of fetuin-A content of CPPs is an exercise of simplification. Given this standpoint, more targeted studies are needed to demonstrate that CPPs are more likely to hold the key on how physiological ossification has correspondence with pathological calcification.

Nevertheless, we need to take into account the current limitations of the T_{50} test. Several weaknesses were identified by Pasch et al. [9]: the test overrides the contribution of VSMCs and calcifying myeloid cells in promoting VC in vivo, and the serum pH had no influence on the test due to the presence of a strong buffer. Another issue is attaining standardized conditions to perform the test. Consequently, even if a reference interval would be preferable to be established, each laboratory is likely to set up its own different reference interval, hence it is hard to envisage an accepted consensus interval. The test is robust but needs further development in terms of time per test which is too long to be clinically implemented for the moment: to perform a 96-well format takes 10 h [9].

In addition, simply minimizing the T_{50} as a marker only for VC may be incomplete. The T_{50} could be considered as a momentary status of the sum of pro- and anti-calcification factors in the serum of a patient, but this may have implications on other pathophysiological processes, thus opening a wide field of research. Accordingly, the term mineral stress has been coined by Pasch et al. [80] and refers to the interaction between inflammation, oxidative stress, and calcification promoted by CPP II.

6. Conclusions

The relatively recent discovery of CPPs opens up new possibilities for the prevention of VC and the attempt to quantify the serum calcification propensity via T_{50}. Even though the factors that influence serum CPP levels, including their ripening process, as well the effect of various factors upon T_{50} and its variation in different diseases is incompletely understood, there is mounting evidence suggesting that T_{50} could be a viable marker in the assessment of VC. Moreover, T_{50} could be valuable in managing VC in CKD patients, especially those undergoing HD, who have a significantly increased risk for developing cardiovascular complications. In these situations, the early introduction of a treatment strategy that increases T_{50} could mitigate the obvious complications related to VC. Such an approach is still at an early phase, warranting future studies on the use of T_{50} as a standard tool in the assessment of VC, thus allowing early measures to prevent cardiovascular complications in patients at risk.

Abbreviations:

BMD	Bone Mineral Density
Ca^{2+}	Calcium
CAC	Coronary Artery Calcification
CKD	Chronic Kidney Disease
CPP I	Primary Calciprotein Particles
CPP II	Secondary Calciprotein Particles
CPP	Calciprotein Particles
CVD	Cardiovascular Disease
CVE	Cardiovascular Event
eGFR	Estimated Glomerular Filtration Rate
ELISA	Enzyme-Linked Immunosorbent Assay
GRP	Gla-Rich Protein
HD	Haemodialysis
H_2S	Hydrogen Sulphide
IL-1β	Interleukin 1β
Mg^{2+}	Magnesium
N/A	Not Applicable
Pi	Phosphate
SLE	Systemic Lupus Erythematosus
tCPP	Total Calciprotein Particles
TNF-α	Tumour Necrosis Factor α
TOD	Time of Death
VC	Vascular Calcification
VSMC	Vascular Smooth Muscle Cell

Author Contributions: All authors have contributed equally to this manuscript. Conceptualization, C.N.S. and T.I.; methodology, T.I.; Writing—original draft, T.I. and C.N.S.; Writing—review and editing, A.J.v.B. and A.M.C. All authors have read and agreed to the published version of the manuscript.

Funding: This work was supported by a grant of the Ministry of Research and Innovation, CNCS—UEFISCDI, project number PN-III-P4-ID-PCCF-2016-0016, within PNCDI III.

Conflicts of Interest: The authors declare no conflict of interest.

References

1. Heiss, A.; Duchesne, A.; Denecke, B.; Grötzinger, J.; Yamamoto, K.; Renné, T.; Jahnen-Dechent, W. Structural Basis of Calcification Inhibition by α2-HS Glycoprotein/Fetuin-A. *J. Boil. Chem.* **2003**, *278*, 13333–13341. [CrossRef]

2. Köppert, S.; Büscher, A.; Babler, A.; Ghallab, A.; Buhl, E.M.; Latz, E.; Hengstler, J.G.; Smith, E.R.; Jahnen-Dechent, W. Cellular Clearance and Biological Activity of Calciprotein Particles Depend on Their Maturation State and Crystallinity. *Front. Immunol.* **2018**, *9*, 9. [CrossRef]

3. Viegas, C.S.; Santos, L.; Macedo, A.; Matos, A.P.; Silva, A.P.; Neves, P.L.; Staes, A.; Gevaert, K.; Morais, R.; Vermeer, C.; et al. Chronic Kidney Disease Circulating Calciprotein Particles and Extracellular Vesicles Promote Vascular Calcification. *Arter. Thromb. Vasc. Boil.* **2018**, *38*, 575–587. [CrossRef]

4. Schinke, T.; Amendt, C.; Trindl, A.; Pöschke, O.; Müller-Esterl, W.; Jahnen-Dechent, W. The Serum Protein α2-HS Glycoprotein/Fetuin Inhibits Apatite Formationin Vitroand in Mineralizing Calvaria Cells. *J. Boil. Chem.* **1996**, *271*, 20789–20796. [CrossRef]

5. Holt, S.G.; Smith, E.R. Fetuin-A-containing calciprotein particles in mineral trafficking and vascular disease. *Nephrol. Dial. Transplant.* **2016**, *31*, 1583–1587. [CrossRef]

6. Heiss, A.; Eckert, T.; Aretz, A.; Richtering, W.; Van Dorp, W.; Schäfer, C.; Jahnen-Dechent, W. Hierarchical Role of Fetuin-A and Acidic Serum Proteins in the Formation and Stabilization of Calcium Phosphate Particles. *J. Boil. Chem.* **2008**, *283*, 14815–14825. [CrossRef]

7. Cai, M.M.X.; Smith, E.R.; Holt, S.G. The role of fetuin-A in mineral trafficking and deposition. *BoneKey Rep.* **2015**, *4*, 672. [CrossRef]

8. Rochette, C.N.; Rosenfeldt, S.; Heiss, A.; Narayanan, T.; Ballauff, M.; Jahnen-Dechent, W. A Shielding Topology Stabilizes the Early Stage Protein-Mineral Complexes of Fetuin-A and Calcium Phosphate: A Time-Resolved Small-Angle X-ray Study. *ChemBioChem* **2009**, *10*, 735–740. [CrossRef]

9. Pasch, A.; Farese, S.; Gräber, S.; Wald, J.; Richtering, W.; Floege, J.; Jahnen-Dechent, W. Nanoparticle-Based Test Measures Overall Propensity for Calcification in Serum. *J. Am. Soc. Nephrol.* **2012**, *23*, 1744–1752. [CrossRef]

10. Hunt, J.L.; Fairman, R.; Mitchell, M.E.; Carpenter, J.P.; Golden, M.; Khalapyan, T.; Wolfe, M.; Neschis, D.; Milner, R.; Scoll, B.; et al. Bone Formation in Carotid Plaques. *Stroke* **2002**, *33*, 1214–1219. [CrossRef]

11. Van Oostrom, O.; Fledderus, J.O.; De Kleijn, D.; Pasterkamp, G.; Verhaar, M. Smooth Muscle Progenitor Cells: Friend or Foe in Vascular Disease? *Curr. Stem Cell Res. Ther.* **2009**, *4*, 131–140. [CrossRef]

12. Persy, V.; D'Haese, P. Vascular calcification and bone disease: The calcification paradox. *Trends Mol. Med.* **2009**, *15*, 405–416. [CrossRef]

13. Ho, C.Y.; Shanahan, C.M. Medial Arterial Calcification. *Arter. Thromb. Vasc. Boil.* **2016**, *36*, 1475–1482. [CrossRef]

14. Lanzer, P.; Boehm, M.; Sorribas, V.; Thiriet, M.; Janzen, J.; Zeller, T.; Hilaire, C.S.; Shanahan, C.M. Medial vascular calcification revisited: Review and perspectives. *Eur. Hear. J.* **2014**, *35*, 1515–1525. [CrossRef]

15. Magne, D.; Julien, M.; Vinatier, C.; Weiss, P.; Guicheux, J.; Merhi-Soussi, F. Cartilage formation in growth plate and arteries: From physiology to pathology. *BioEssays* **2005**, *27*, 708–716. [CrossRef]

16. Goodman, W.G.; Goldin, J.; Kuizon, B.D.; Yoon, C.; Gales, B.; Sider, D.; Wang, Y.; Chung, J.; Emerick, A.; Greaser, L.; et al. Coronary-Artery Calcification in Young Adults with End-Stage Renal Disease Who Are Undergoing Dialysis. *New Engl. J. Med.* **2000**, *342*, 1478–1483. [CrossRef]

17. Bressendorff, I.; Hansen, D.; Pasch, A.; Holt, S.G.; Schou, M.; Brandi, L.; Smith, E.R. The effect of increasing dialysate magnesium on calciprotein particles, inflammation and bone markers: Post hoc analysis from a randomized controlled clinical trial. *Nephrol. Dial. Transplant.* **2019**. [CrossRef]

18. Shishkova, D.; Velikanova, E.; Sinitsky, M.; Tsepokina, A.; Gruzdeva, O.V.; Bogdanov, L.; Kutikhin, A. Calcium Phosphate Bions Cause Intimal Hyperplasia in Intact Aortas of Normolipidemic Rats through Endothelial Injury. *Int. J. Mol. Sci.* **2019**, *20*, 5728. [CrossRef]

19. Ter Braake, A.D.; Eelderink, C.; Zeper, L.W.; Pasch, A.; Bakker, S.J.L.; De Borst, M.H.; Hoenderop, J.G.J.; De Baaij, J.H. Calciprotein particle inhibition explains magnesium-mediated protection against vascular calcification. *Nephrol. Dial. Transplant.* **2019**. [CrossRef]

20. Nakazato, J.; Hoshide, S.; Wake, M.; Miura, Y.; Kuro-O, M.; Kario, K. Association of calciprotein particles measured by a new method with coronary artery plaque in patients with coronary artery disease: A cross-sectional study. *J. Cardiol.* **2019**, *74*, 428–435. [CrossRef]

21. Nakamura, K.; Nagata, Y.; Hiroyoshi, T.; Isoyama, N.; Fujikawa, K.; Miura, Y.; Matsuyama, H.; Kuro-O, M. The effect of lanthanum carbonate on calciprotein particles in hemodialysis patients. *Clin. Exp. Nephrol.* **2019**, *24*, 323–329. [CrossRef]

22. Nemoto, Y.; Kumagai, T.; Ishizawa, K.; Miura, Y.; Shiraishi, T.; Morimoto, C.; Sakai, K.; Omizo, H.; Yamazaki, O.; Tamura, Y.; et al. Phosphate binding by sucroferric oxyhydroxide ameliorates renal injury in the remnant kidney model. *Sci. Rep.* **2019**, *9*, 1732. [CrossRef]

23. Chen, W.; Anokhina, V.; Dieudonne, G.; Abramowitz, M.K.; Kashyap, R.; Yan, C.; Wu, T.T.; Bentley, K.L.D.M.; Miller, B.L.; Bushinsky, D. Patients with advanced chronic kidney disease and vascular calcification have a large hydrodynamic radius of secondary calciprotein particles. *Nephrol. Dial. Transplant.* **2019**, *34*, 992–1000. [CrossRef]

24. Cai, M.M.X.; Smith, E.R.; Kent, A.; Huang, L.; Hewitson, T.; McMahon, L.P.; Holt, S.G. Calciprotein Particle Formation in Peritoneal Dialysis Effluent is Dependent on Dialysate Calcium Concentration. *Perit. Dial. Int.* **2018**, *38*, 286–292. [CrossRef]

25. Ruderman, I.; Smith, E.R.; Toussaint, N.D.; Hewitson, T.; Holt, S.G. Longitudinal changes in bone and mineral metabolism after cessation of cinacalcet in dialysis patients with secondary hyperparathyroidism. *BMC Nephrol.* **2018**, *19*, 113. [CrossRef]

26. Aghagolzadeh, P.; Radpour, R.; Bachtler, M.; Van Goor, H.; Smith, E.R.; Lister, A.; Odermatt, A.; Feelisch, M.; Pasch, A. Hydrogen sulfide attenuates calcification of vascular smooth muscle cells via KEAP1/NRF2/NQO1 activation. *Atherosclerosis* **2017**, *265*, 78–86. [CrossRef]

27. Cai, M.M.X.; Smith, E.R.; Tan, S.-J.; Hewitson, T.; Holt, S.G. The Role of Secondary Calciprotein Particles in the Mineralisation Paradox of Chronic Kidney Disease. *Calcif. Tissue Int.* **2017**, *101*, 570–580. [CrossRef]

28. Yamada, H.; Kuro-O, M.; Ishikawa, S.-E.; Funazaki, S.; Kusaka, I.; Kakei, M.; Hara, K. Daily variability in serum levels of calciprotein particles and their association with mineral metabolism parameters: A cross-sectional pilot study. *Nephrology* **2018**, *23*, 226–230. [CrossRef]

29. Aghagolzadeh, P.; Bachtler, M.; Bijarnia, R.; Jackson, C.B.; Smith, E.R.; Odermatt, A.; Radpour, R.; Pasch, A. Calcification of vascular smooth muscle cells is induced by secondary calciprotein particles and enhanced by tumor necrosis factor-α. *Atherosclerosis* **2016**, *251*, 404–414. [CrossRef]

30. Cai, M.M.X.; Wigg, B.; Smith, E.R.; Hewitson, T.; McMahon, L.P.; Holt, S.G. Relative abundance of fetuin- A in peritoneal dialysis effluent and its association with in situ formation of calciprotein particles: An observational pilot study. *Nephrology* **2014**, *20*, 6–10. [CrossRef]

31. Smith, E.R.; Hanssen, E.; McMahon, L.P.; Holt, S.G. Fetuin-A-Containing Calciprotein Particles Reduce Mineral Stress in the Macrophage. *PLoS ONE* **2013**, *8*, e60904. [CrossRef] [PubMed]

32. Smith, E.R.; Cai, M.M.; McMahon, L.P.; Pedagogos, E.; Toussaint, N.D.; Brumby, C.; Holt, S.G. Serum fetuin-A concentration and fetuin-A-containing calciprotein particles in patients with chronic inflammatory disease and renal failure. *Nephrology* **2013**, *18*, 215–221. [CrossRef] [PubMed]

33. Smith, E.R.; Ford, M.L.; Tomlinson, L.; Rajkumar, C.; McMahon, L.P.; Holt, S.G. Phosphorylated fetuin-A-containing calciprotein particles are associated with aortic stiffness and a procalcific milieu in patients with pre-dialysis CKD. *Nephrol. Dial. Transplant.* **2012**, *27*, 1957–1966. [CrossRef] [PubMed]

34. Ter Meulen, K.J.; Dekker, M.J.E.; Pasch, A.; Broers, N.J.H.; Van Der Sande, F.M.; Kooman, J.P.; Konings, C.J.A.M.; Gsponer, I.M.; Bachtler, M.D.N.; Gauly, A.; et al. Citric-acid dialysate improves the calcification propensity of hemodialysis patients: A multicenter prospective randomized cross-over trial. *PLoS ONE* **2019**, *14*, e0225824. [CrossRef]

35. Van Dijk, P.R.; Hop, H.; Waanders, F.; Mulder, U.J.; Pasch, A.; Hillebrands, J.-L.; Van Goor, H.; Bilo, H.J. Serum calcification propensity in type 1 diabetes associates with mineral stress. *Diabetes Res. Clin. Pract.* **2019**, *158*, 107917. [CrossRef]

36. Bundy, J.D.; Cai, X.; Mehta, R.C.; Scialla, J.J.; De Boer, I.H.; Hsu, C.-Y.; Go, A.S.; Dobre, M.; Chen, J.; Rao, P.S.; et al. Serum Calcification Propensity and Clinical Events in CKD. *Clin. J. Am. Soc. Nephrol.* **2019**, *14*, 1562–1571. [CrossRef]

37. Ponte, B.; Pruijm, M.; Pasch, A.; Dufey-Teso, A.; Martin, P.-Y.; De Seigneux, S. Dialysis initiation improves calcification propensity. *Nephrol. Dial. Transplant.* **2019**, *35*, 495–502. [CrossRef]

38. Bundy, J.D.; Cai, X.; Scialla, J.J.; Dobre, M.A.; Chen, J.; Hsu, C.-Y.; Leonard, M.B.; Go, A.S.; Rao, P.S.; Lash, J.P.; et al. Serum Calcification Propensity and Coronary Artery Calcification Among Patients With CKD: The CRIC (Chronic Renal Insufficiency Cohort) Study. *Am. J. Kidney Dis.* **2019**, *73*, 806–814. [CrossRef]

39. Quiñones, H.; Hamdi, T.; Sakhaee, K.; Pasch, A.; Moe, O.W.; Pak, C.Y.C. Control of metabolic predisposition to cardiovascular complications of chronic kidney disease by effervescent calcium magnesium citrate: A feasibility study. *J. Nephrol.* **2018**, *32*, 93–100. [CrossRef]

40. Aigner, C.; Cejka, D.; Sliber, C.; Fraunschiel, M.; Sunder-Plassmann, G.; Gaggl, M. Oral Sodium Bicarbonate Supplementation Does Not Affect Serum Calcification Propensity in Patients with Chronic Kidney Disease and Chronic Metabolic Acidosis. *Kidney Blood Press. Res.* **2019**, *44*, 188–199. [CrossRef]

41. Bullen, A.L.; Anderson, C.A.M.; Hooker, E.R.; Kado, D.M.; Orwoll, E.; Pasch, A.; Ix, J.H. Correlates of T50 and relationships with bone mineral density in community-living older men: The osteoporotic fractures in men (MrOS) study. *Osteoporos. Int.* **2019**, *30*, 1529–1531. [CrossRef] [PubMed]

42. Bostom, A.; Pasch, A.; Madsen, T.; Roberts, M.B.; Franceschini, N.; Steubl, D.; Garimella, P.S.; Ix, J.H.; Tuttle, K.R.; Ivanova, A.; et al. Serum Calcification Propensity and Fetuin-A: Biomarkers of Cardiovascular Disease in Kidney Transplant Recipients. *Am. J. Nephrol.* **2018**, *48*, 21–31. [CrossRef] [PubMed]

43. Andrews, E.S.; Perrenoud, L.; Nowak, K.L.; You, Z.; Pasch, A.; Chonchol, M.; Kendrick, J.; Jalal, D. Examining the effects of uric acid-lowering on markers vascular of calcification and CKD-MBD. A post-hoc analysis of a randomized clinical trial. *PLoS ONE* **2018**, *13*, e0205831. [CrossRef] [PubMed]

44. Ussif, A.M.; Pihlstrøm, H.; Pasch, A.; Holdaas, H.; Hartmann, A.; Smerud, K.; Åsberg, A. Paricalcitol supplementation during the first year after kidney transplantation does not affect calcification propensity score. *BMC Nephrol.* **2018**, *19*, 212. [CrossRef]

45. Bressendorff, I.; Hansen, D.; Schou, M.; Pasch, A.; Brandi, L. The Effect of Increasing Dialysate Magnesium on Serum Calcification Propensity in Subjects with End Stage Kidney Disease. *Clin. J. Am. Soc. Nephrol.* **2018**, *13*, 1373–1380. [CrossRef]

46. Lorenz, G.; Mayer, C.C.; Bachmann, Q.; Stryeck, S.; Braunisch, M.C.; Haller, B.; Carbajo-Lozoya, J.; Schmidt, A.; Witthauer, S.; Abuzahu, J.; et al. Acetate-free, citrate-acidified bicarbonate dialysis improves serum calcification propensity—a preliminary study. *Nephrol. Dial. Transplant.* **2018**, *33*, 2043–2051. [CrossRef]

47. Dahdal, S.; Devetzis, V.; Chalikias, G.; Tziakas, D.; Chizzolini, C.; Ribi, C.; Trendelenburg, M.; Eisenberger, U.; Hauser, T.; Pasch, A.; et al. Serum calcification propensity is independently associated with disease activity in systemic lupus erythematosus. *PLoS ONE* **2018**, *13*, e0188695. [CrossRef]

48. Pruijm, M.; Lu, Y.; Megdiche, F.; Piskunowicz, M.; Milani, B.; Stuber, M.; Bachtler, M.; Vogt, B.; Burnier, M.; Pasch, A. Serum calcification propensity is associated with renal tissue oxygenation and resistive index in patients with arterial hypertension or chronic kidney disease. *J. Hypertens.* **2017**, *35*, 2044–2052. [CrossRef]

49. Pasch, A.; Block, G.A.; Bachtler, M.; Smith, E.R.; Jahnen-Dechent, W.; Arampatzis, S.; Chertow, G.M.; Parfrey, P.; Ma, X.; Floege, J. Blood Calcification Propensity, Cardiovascular Events, and Survival in Patients Receiving Hemodialysis in the EVOLVE Trial. *Clin. J. Am. Soc. Nephrol.* **2016**, *12*, 315–322. [CrossRef]

50. Bielesz, B.; Reiter, T.; Marculescu, R.; Gleiss, A.; Bojic, M.; Kieweg, H.; Cejka, D. Calcification Propensity of Serum is Independent of Excretory Renal Function. *Sci. Rep.* **2017**, *7*, 17941. [CrossRef]

51. Bressendorff, I.; Hansen, D.; Schou, M.; Silver, B.; Pasch, A.; Bouchelouche, P.; Pedersen, L.; Rasmussen, L.M.; Brandi, L. Oral Magnesium Supplementation in Chronic Kidney Disease Stages 3 and 4: Efficacy, Safety, and Effect on Serum Calcification Propensity-A Prospective Randomized Double-Blinded Placebo-Controlled Clinical Trial. *Kidney Int. Rep.* **2016**, *2*, 380–389. [CrossRef] [PubMed]

52. Lorenz, G.; Steubl, D.; Kemmner, S.; Pasch, A.; Koch-Sembdner, W.; Pham, D.; Haller, B.; Bachmann, Q.; Mayer, C.C.; Wassertheurer, S.; et al. Worsening calcification propensity precedes all-cause and cardiovascular mortality in haemodialyzed patients. *Sci. Rep.* **2017**, *7*, 13368. [CrossRef] [PubMed]

53. Smerud, K.; Åsberg, A.; Kile, H.; Pasch, A.; Dahle, D.O.; Bollerslev, J.; Godang, K.; Hartmann, A. A rapid and sustained improvement of calcification propensity score (serum T50) after successful kidney transplantation: Reanalysis of a randomized controlled trial of ibandronate. *Clin. Transplant.* **2017**, *31*, e13131. [CrossRef] [PubMed]

54. Dekker, M.; Pasch, A.; Van Der Sande, F.; Konings, C.; Bachtler, M.; Dionisi, M.; Meier, M.; Kooman, J.; Canaud, B. High-Flux Hemodialysis and High-Volume Hemodiafiltration Improve Serum Calcification Propensity. *PLoS ONE* **2016**, *11*, e0151508. [CrossRef] [PubMed]

55. Bristow, S.; Gamble, G.D.; Pasch, A.; O'Neill, W.C.; Stewart, A.; Horne, A.; Reid, I.R. Acute and 3-month effects of calcium carbonate on the calcification propensity of serum and regulators of vascular calcification: Secondary analysis of a randomized controlled trial. *Osteoporos. Int.* **2015**, *27*, 1209–1216. [CrossRef] [PubMed]

56. Dahle, D.O.; Åsberg, A.; Hartmann, A.; Holdaas, H.; Bachtler, M.; Jenssen, T.G.; Dionisi, M.; Pasch, A. Serum Calcification Propensity Is a Strong and Independent Determinant of Cardiac and All-Cause Mortality in Kidney Transplant Recipients. *Arab. Archaeol. Epigr.* **2015**, *16*, 204–212. [CrossRef]

57. Keyzer, C.A.; De Borst, M.H.; Berg, E.V.D.; Jahnen-Dechent, W.; Arampatzis, S.; Farese, S.; Bergmann, I.P.; Floege, J.; Navis, G.; Bakker, S.J.; et al. Calcification Propensity and Survival among Renal Transplant Recipients. *J. Am. Soc. Nephrol.* **2015**, *27*, 239–248. [CrossRef]

58. Berchtold, L.; Ponte, B.; Moll, S.; Hadaya, K.; Seyde, O.; Bachtler, M.; Vallée, J.-P.; Martin, P.-Y.; Pasch, A.; De Seigneux, S. Phosphocalcic Markers and Calcification Propensity for Assessment of Interstitial Fibrosis and Vascular Lesions in Kidney Allograft Recipients. *PLoS ONE* **2016**, *11*, e0167929. [CrossRef]

59. De Seigneux, S.; Ponte, B.; Berchtold, L.; Hadaya, K.; Martin, P.-Y.; Pasch, A. Living kidney donation does not adversely affect serum calcification propensity and markers of vascular stiffness. *Transpl. Int.* **2015**, *28*, 1074–1080. [CrossRef]

60. Smith, E.R.; Ford, M.L.; Tomlinson, L.; Bodenham, E.; McMahon, L.P.; Farese, S.; Rajkumar, C.; Holt, S.G.; Pasch, A. Serum Calcification Propensity Predicts All-Cause Mortality in Predialysis CKD. *J. Am. Soc. Nephrol.* **2013**, *25*, 339–348. [CrossRef]

61. Zarb, Y.; Weber-Stadlbauer, U.; Kirschenbaum, D.; Kindler, D.R.; Richetto, J.; Keller, D.; Rademakers, R.; Dickson, D.W.; Pasch, A.; Byzova, T.V.; et al. Ossified blood vessels in primary familial brain calcification elicit a neurotoxic astrocyte response. *Brain* **2019**, *142*, 885–902. [CrossRef] [PubMed]

62. Kendrick, J.; Shah, P.; Andrews, E.; You, Z.; Nowak, K.L.; Pasch, A.; Chonchol, M. Effect of Treatment of Metabolic Acidosis on Vascular Endothelial Function in Patients with CKD. *Clin. J. Am. Soc. Nephrol.* **2018**, *13*, 1463–1470. [CrossRef] [PubMed]

63. Voelkl, J.; Tuffaha, R.; Luong, T.T.; Zickler, D.; Masyout, J.; Feger, M.; Verheyen, N.; Blaschke, F.; Kuro-O, M.; Tomaschitz, A.; et al. Zinc Inhibits Phosphate-Induced Vascular Calcification through TNFAIP3-Mediated Suppression of NF-κB. *J. Am. Soc. Nephrol.* **2018**, *29*, 1636–1648. [CrossRef] [PubMed]

64. Jahnen-Dechent, W.; Trindl, A.; Godovac-Zimmermann, J.; Müller-Esterl, W. Posttranslational Processing of Human alpha2-HS Glycoprotein (Human Fetuin). Evidence for the Production of a Phosphorylated Single-Chain Form by Hepatoma Cells. *JBIC J. Boil. Inorg. Chem.* **1994**, *226*, 59–69. [CrossRef] [PubMed]

65. Mathews, S.T.; Chellam, N.; Srinivas, P.R.; Cintron, V.J.; Leon, M.; Goustin, A.S.; Grunberger, G. Alpha2-HSG, a specific inhibitor of insulin receptor autophosphorylation, interacts with the insulin receptor. *Mol. Cell. Endocrinol.* **2000**, *164*, 87–98. [CrossRef]

66. Auberger, P.; Falquerho, L.; Contreres, J.O.; Pagès, G.; Le Cam, G.; Rossi, B.; Le Cam, A. Characterization of a natural inhibitor of the insulin receptor tyrosine kinase: cDNA cloning, purification, and anti-mitogenic activity. *Cell* **1989**, *58*, 631–640. [CrossRef]

67. Suzuki, M.; Shimokawa, H.; Takagi, Y.; Sasaki, S. Calcium-binding properties of fetuin in fetal bovine serum. *J. Exp. Zoöl.* **1994**, *270*, 501–507. [CrossRef]

68. Herrmann, M.; Schäfer, C.; Heiss, A.; Gräber, S.; Kinkeldey, A.; Büscher, A.; Schmitt, M.M.; Bornemann, J.; Nimmerjahn, F.; Herrmann, M.; et al. Clearance of Fetuin-A–Containing Calciprotein Particles Is Mediated by Scavenger Receptor-A. *Circ. Res.* **2012**, *111*, 575–584. [CrossRef]

69. Hamano, T.; Matsui, I.; Mikami, S.; Tomida, K.; Fujii, N.; Imai, E.; Rakugi, H.; Isaka, Y. Fetuin-mineral complex reflects extraosseous calcification stress in CKD. *J. Am. Soc. Nephrol.* **2010**, *21*, 1998–2007. [CrossRef]
70. Smith, E.R.; Hewitson, T.; Cai, M.M.X.; Aghagolzadeh, P.; Bachtler, M.; Pasch, A.; Holt, S.G. A novel fluorescent probe-based flow cytometric assay for mineral-containing nanoparticles in serum. *Sci. Rep.* **2017**, *7*, 5686. [CrossRef]
71. Ismail, A.H.; Schäfer, C.; Heiss, A.; Walter, M.; Jahnen-Dechent, W.; Leonhardt, S. An electrochemical impedance spectroscopy (EIS) assay measuring the calcification inhibition capacity in biological fluids. *Biosens. Bioelectron.* **2011**, *26*, 4702–4707. [CrossRef] [PubMed]
72. Ter Braake, A.D.; Tinnemans, P.T.; Shanahan, C.M.; Hoenderop, J.G.J.; De Baaij, J.H. Magnesium prevents vascular calcification in vitro by inhibition of hydroxyapatite crystal formation. *Sci. Rep.* **2018**, *8*, 2069. [CrossRef] [PubMed]
73. Huang, M.; Zheng, L.; Xu, H.; Tang, D.; Lin, L.; Zhang, J.; Li, C.; Wang, W.; Yuan, Q.; Tao, L.; et al. Oxidative stress contributes to vascular calcification in patients with chronic kidney disease. *J. Mol. Cell. Cardiol.* **2020**, *138*, 256–268. [CrossRef] [PubMed]
74. Rubin, M.R.; Silverberg, S.J. Vascular calcification and osteoporosis–the nature of the nexus. *J. Clin. Endocrinol. Metab.* **2004**, *89*, 4243–4245. [CrossRef] [PubMed]
75. Keeley, F.; Sitarz, E. Identification and quantitation of α2-HS-glycoprotein in the mineralized matrix of calcified plaques of atherosclerotic human aorta. *Atherosclerosis* **1985**, *55*, 63–69. [CrossRef]
76. Binkert, C.; Demetriou, M.; Sukhu, B.; Szweras, M.; Tenenbaum, H.C.; Dennis, J. Regulation of osteogenesis by fetuin. *J. Boil. Chem.* **1999**, *274*, 28514–28520. [CrossRef]
77. Ketteler, M.; Bongartz, P.; Westenfeld, R.; Wildberger, J.E.; Mahnken, A.H.; Böhm, R.; Metzger, T.; Wanner, C.; Jahnen-Dechent, W.; Floege, J. Association of low fetuin-A (AHSG) concentrations in serum with cardiovascular mortality in patients on dialysis: A cross-sectional study. *Lancet* **2003**, *361*, 827–833. [CrossRef]
78. Fontaine, M.A.; Albert, A.; Dubois, B.; Saint-Remy, A.; Rorive, G. Fracture and bone mineral density in hemodialysis patients. *Clin. Nephrol.* **2000**, *54*, 218–226.
79. Kirkpantur, A.; Altun, B.; Hazirolan, T.; Akata, D.; Arici, M.; Kirazli, S.; Turgan, C. Association Among Serum Fetuin-A Level, Coronary Artery Calcification, and Bone Mineral Densitometry in Maintenance Hemodialysis Patients. *Artif. Organs* **2009**, *33*, 844–854. [CrossRef]
80. Pasch, A.; Jahnen-Dechent, W.; Smith, E.R. Phosphate, Calcification in Blood, and Mineral Stress: The Physiologic Blood Mineral Buffering System and Its Association with Cardiovascular Risk. *Int. J. Nephrol.* **2018**, *2018*, 1–5. [CrossRef]

Journal of
Clinical Medicine

Review

New Treatment Options for Hyperkalemia in Patients with Chronic Kidney Disease

Pasquale Esposito *, Novella Evelina Conti, Valeria Falqui, Leda Cipriani, Daniela Picciotto, Francesca Costigliolo, Giacomo Garibotto, Michela Saio and Francesca Viazzi

Clinica Nefrologica, Dialisi, Trapianto, Department of Internal Medicine, University of Genoa and IRCCS Ospedale Policlinico San Martino, Viale Benedetto XV, 16132 Genoa, Italy; novella.conti@hsanmartino.it (N.E.C.); valeria.falqui@hsanmartino.it (V.F.); cipriani.leda@gmail.com (L.C.); danipicciotto@me.com (D.P.); 3232594@studenti.unige.it (F.C.); gari@unige.it (G.G.); michela.saio@virgilio.it (M.S.); francesca.viazzi@unige.it (F.V.)
* Correspondence: pasqualeesposito@hotmail.com; Tel.: +39-0105-5553-62

Received: 3 June 2020; Accepted: 21 July 2020; Published: 22 July 2020

Abstract: Hyperkalemia may cause life-threatening cardiac and neuromuscular alterations, and it is associated with high mortality rates. Its treatment includes a multifaceted approach, guided by potassium levels and clinical presentation. In general, treatment of hyperkalemia may be directed towards stabilizing cell membrane potential, promoting transcellular potassium shift and lowering total K^+ body content. The latter can be obtained by dialysis, or by increasing potassium elimination by urine or the gastrointestinal tract. Until recently, the only therapeutic option for increasing fecal K^+ excretion was represented by the cation-exchanging resin sodium polystyrene sulfonate. However, despite its common use, the efficacy of this drug has been poorly studied in controlled studies, and concerns about its safety have been reported. Interestingly, new drugs, namely patiromer and sodium zirconium cyclosilicate, have been developed to treat hyperkalemia by increasing gastrointestinal potassium elimination. These medications have proved their efficacy and safety in large clinical trials, involving subjects at high risk of hyperkalemia, such as patients with heart failure and chronic kidney disease. In this review, we discuss the mechanisms of action and the updated data of patiromer and sodium zirconium cyclosilicate, considering that the availability of these new treatment options offers the possibility of improving the management of both acute and chronic hyperkalemia.

Keywords: hyperkalemia; chronic kidney disease; heart failure; sodium polystyrene sulfonate; patiromer; sodium zirconium cyclosilicate

---

1. Introduction

Potassium (K^+) is a key element in body physiology. It regulates many biological processes, such as acid–base homeostasis, hormone secretion, systemic blood pressure control and gastrointestinal motility [1]. However, probably the most important role of K^+ is its participation in generating bioelectricity, by establishing ion gradients and flows between the extracellular and intracellular spaces, thus regulating resting membrane potential and cellular excitability, which are essential to the function of excitable tissues, such as nerve, muscle and cardiac conduction tissues.

This function is a consequence of the high compartmentalization of K^+, due to the ubiquitous presence of plasma membrane Na-K-ATPases, which pump sodium out of, and K^+ into, the cell [2]. Therefore, K^+ results from the most concentrated intracellular electrolyte, while its extracellular concentration is extremely low. We estimate a total K^+ body content of approximately 50 mEq/ kg (i.e., 3500 mEq in a 70-kg person); about 98% of this K^+ is within cells, while only 2% (70 mEq) is in the extracellular fluid, where it reaches normal concentrations of 3.5 to 5.0 mmol/L [3].

Hyperkalemia is defined as a serum potassium level greater than 5.0 mmol/L, while severe hyperkalemia is defined as a level greater than 6.0 mmol/L. It is a very common disorder, and in the United States more than 800,000 emergency department (ED) visits occur annually because of hyperkalemia [4]. The actual incidence and prevalence of hyperkalemia in the general population are unknown, but studies based on large cohorts have reported incidence rates between 1 and 3 per 100 persons per year, rising to 10% in hospitalized patients [5,6].

Moreover, hyperkalemia prevalence may be significantly high in the presence of certain predisposing conditions. So, although the available data are not uniform [7,8], an analysis of a large geographically diverse population showed basal potassium values of $\geq$ 5.0 mmol/L in 9.1% of patients with chronic heart failure (CHF), in 11.5% of chronic kidney disease (CKD) stage 3–5 patients, in 8.3% of patients with diabetes, and in 13.1% of those patients with all these conditions [9]. In addition, among CKD patients, those requiring dialysis represent a group at particularly high risk of hyperkalemia [10].

Clinical complications and death in hyperkalemia patients are mainly determined by the cardiac electrophysiological effects of elevated K^+ levels [11]. Indeed, hyperkalemia, by diminishing the K^+ intracellular/K^+ extracellular ratio, reduces the membrane potential, causing a partial depolarization of the cell membrane, which results in an initial increase in conduction velocity. Then, if persistent and profound, hyperkalemia also decreases membrane excitability by the inactivation of the voltage-gated sodium channels, making the cell refractory to excitation, and thus leading to arrhythmias and heart block [12]. Moreover, besides cardiac effects, hyperkalemia can also cause other physiologic perturbations, such as muscle weakness progressing to flaccid paralysis, and metabolic acidosis, which in turn may contribute to the progression of CKD [13].

The treatment of hyperkalemia may involve the recognition different time-points and goals, guided by potassium levels and the severity of the clinical presentation. In general, the first aim is to prevent cardiac consequences and lower serum potassium to safe levels as soon as possible; then it is important to reduce the K^+ body content, aiming to maintain serum potassium at normal values [14]. The latter can be obtained by dialysis, or by increasing potassium elimination via urine or the gastrointestinal tract. For a long time, the only therapeutic option for increasing fecal K^+ excretion has been represented by sodium polystyrene sulfonate, a cation-exchanging resin the efficacy and safety of which have been questioned. Recently, new drugs able to promote gastrointestinal potassium elimination, namely patiromer and sodium zirconium cyclosilicate, have been developed and studied in large trials, proving their efficacy and safety in different clinical contexts. In this review, we briefly discuss the pathophysiology of potassium homeostasis and hyperkalemia, focusing attention on the mechanisms of action and the clinical data of patiromer and sodium zirconium cyclosilicate, considering that these new treatments may represent a chance to improve the management of both acute and chronic hyperkalemia.

2. Potassium Homeostasis: An Overview

Due to its important functions, and considering that large deviations in K^+ serum levels are not compatible with life, K^+ homeostasis is finely regulated by numerous mechanisms.

Classically, we distinguish between external and internal K^+ balance [3]. External K^+ balance regulates K^+ body content, and it is the result of the relationship between K^+ assumption (by diet or other sources, such as infusions) and K^+ excretion, which is a function of the kidney and the gut. Internal balance accounts for K^+ distribution across cell compartments, which can be influenced by several factors and may be important in determining the actual K^+ extracellular level.

The external potassium balance is mainly influenced by K^+ excretion via the kidney. The normal kidney has a large capacity to excrete potassium and maintain a normal serum potassium concentration. Potassium is freely filtrated by the glomerulus, and is then reabsorbed by the proximal tubule and thick ascending limb, such that only a small amount reaches the aldosterone-sensitive distal nephron, where K^+ excretion, coupled with sodium reabsorption, is finely regulated [15].

The main factors modulating renal K^+ excretion are sodium delivery to the distal nephron, K^+ serum levels and aldosterone plasma concentration. In particular, a relevant role in regulating K^+ homeostasis is played by the adrenal glands, where aldosterone is synthesized, in a negative feedback loop in response to high K^+ levels [16].

Interestingly, evidence is emerging concerning the role of the central nervous system in influencing circadian variability in relation to potassium excretion [1]. Apart from the kidneys, the gastrointestinal tract also contributes to K^+ excretion. In healthy subjects, this contribution is minimal (about 10% of the total), while in the case of renal disease it may increase until it accounts for 50% of the total potassium excretion in patients on dialysis [17]. However, these systems are strictly related, and recently it has been shown that the K^+ enteral load may influence renal excretion, suggesting the presence of a gut-dependent kaluresis, the mechanisms of which are still under investigation [18].

Complementarily to the external K^+ balance, the internal mechanisms of K^+ distribution are very important in regulating K^+ homeostasis and extracellular levels. The physiological factors involved in modulating the shifting of potassium into the cells include the acid–base balance, insulin, and beta-adrenergic stimulation [19]. In particular, metabolic acidosis induces K^+ shift from the intra-to the extracellular space, while the opposite is mediated by insulin and beta-adrenergic signaling. However, external and internal K^+ regulatory mechanisms are integrated, and need to always be active in order to maintain K^+ homeostasis.

A western diet typically contains approximately 50–100 mEq (2–4 g), while a potassium intake of about 90–120 mEq/day (3.5–4.5 g) is recommended [20]. Consequently, the potassium content in a meal may be higher than the potassium present in the plasma, so compensatory mechanisms are necessary in order to avoid a rapid rise in extracellular K^+ levels. For example, after a potassium-rich meal, the K^+ transcellular shift into the cells is suddenly activated, until the kidney reestablishes total body potassium content trough the adjustment of renal potassium excretion.

Knowledge of the basal physiologic regulators of external and internal K^+ homeostasis is necessary in order to understand the main clinical conditions leading to K^+ dysregulation and the appropriate therapeutic approaches, providing, at the same time, a rational basis for developing new drugs [21].

3. Hyperkalemia: Physiopathology, Risk Factors, Clinical Consequences

Hyperkalemia may be caused by several conditions that may alter K^+ homeostasis [1]. First, it could be the consequence of an increased K^+ body content due to excessive K^+ intake, or, more commonly, due to reduced renal excretion. Renal K^+ excretion may be impaired as a result of advanced renal damage. Indeed, while the normal kidney presents adaptation mechanisms that preserve potassium homeostasis, the diseased kidney has a much lower capacity for handling acute potassium loads [22]. Failures of the kidneys in regulating the potassium balance may result from multiple factors, including a reduced glomerular filtration rate, decreased distal delivery of sodium, intrinsic abnormalities of the distal nephron, and decreased mineralocorticoid activity (e.g., hypoaldosteronism), which impair the capacity of the distal nephron to eliminate K^+ from the urine. Moreover, concomitant metabolic alterations, such as acidemia and hyperglycemia, may also concur [23].

Hypoaldosteronism, in turn, may be caused by diabetes, adrenal disease, numerous drugs (e.g., nonsteroidal anti-inflammatory drugs, beta-blockers, inhibitors of the renin-angiotensin-aldosterone system-RAASi, mineralocorticoid receptor blockers, calcineurin-inhibitors, etc.) and old age [24]. Beyond an increase of K^+ body content, alterations in the K^+ distribution across cell compartments can also lead to hyperkalemia. These conditions determine the net release of potassium from damaged cells, such as in cases of trauma, rhabdomyolysis or hemolysis. Moreover, an impaired distribution of K^+ between the intracellular and extracellular spaces can also be due to metabolic acidosis, decompensated diabetes or dysfunctions of the autonomic nervous system.

Considering the physiopathology of potassium homeostasis, it is not surprising that advanced age, chronic kidney disease (CKD), chronic heart failure (CHF), diabetes, use of RAASi (such as

ACE-inhibitors-ACEi and angiotensin receptor blockers-ARB) and mineralocorticoid receptor blockers (MRA) constitute the main risk factors in the development of hyperkalemia [9].

However, a special consideration must be given to the risk of hyperkalemia linked to the use of RAASi and MRA. Indeed, these drugs, because of the evidence of their morbidity and mortality benefits, are widely prescribed to fragile patients, such as patients with diabetes, CKD and CHF [25]. Several studies have evaluated the risk of hyperkalemia associated with RAASi therapy. For example, in the Stockholm Creatinine Measurements (SCREAM) project, 69,426 new users of ACEi/ARB therapy were followed for one year. Overall, hyperkalemia occurred in 1.7% of the entire cohort, but its incidence rose to 29% in patients with severe CKD [26]. Moreover, the risk of hyperkalemia seems further increased when combined RAASi therapy is prescribed. In the Veterans Affairs Nephropathy in Diabetes (VA NEPHRON-D) study, designed to assess the safety of RAASi for type II diabetic kidney disease patients, hyperkalemia was observed in 18.4% of the patients on losartan monotherapy, and in 31.5% of those on a combination therapy with lisinopril [27].

Following the results of these and other studies, the combination of ACEi and ARB is no longer recommended [28]. Similar results have been observed among patients treated with the combination of RAASi and MRA, which, although effective in improving clinical outcomes, may expose patients to a high risk of hyperkalemia [25,29].

The early recognition and treatment of hyperkalemia is essential, because this condition, although often clinically silent, may have severe consequences. Hyperkalemia is associated with poor outcomes and high mortality rates, both in the general population and in different clinical settings, including patients with cardiac and renal diseases and critically ill patients [30,31]. Moreover, hyperkalemia has been described as an independent predictor of mortality in patients admitted to the ED [32].

4. Hyperkalemia: Treatment Strategies

From the pathophysiological point of view, the therapeutic approaches to hyperkalemia can have three different targets: (i) cell membrane potential stabilization; (ii) shifting potassium from extracellular spaces into the cells (i.e., acting on internal K^+ balance); and (iii) lowering K^+ levels and enhancing potassium elimination (i.e., acting on external K^+ balance).

Membrane stabilization may be achieved through the administration of intravenous calcium (calcium chloride or calcium gluconate), while potassium redistribution may be promoted using insulin/glucose, beta-adrenergic agonists (such as albuterol and salbutamol, both intravenous and inhaled) and sodium bicarbonate [33]. These treatments are often preferred in emergency interventions, since they can reduce K^+ levels within a few minutes [34]. However, while they act rapidly, their effects also fade very rapidly.

So, complementary to the strategies that promote the shifting of potassium into cells, the reestablishment of potassium homeostasis should include the reduction of body K^+ content.

This can be achieved through the limitation of potassium intake and the use of medications that increase potassium elimination via urine or the gastrointestinal tract (GI), such as loop diuretics or cation-exchanging resins, or alternatively, by use of hemodialysis, which can reduce body K^+ content but usually require more time to act [35].

In particular, the use of drugs increasing GI potassium elimination is valuable in patients with advanced CKD, who, as discussed above, present significant fecal K^+ excretion.

Although these treatments are widely used in clinical practice, it should be recognized that, as also shown by a recent Cochrane review, standardized therapeutic protocols do not exist [36]. So, in a prospective multicenter study exploring real-life hyperkalemia management in many EDs, it has been shown that there was a great heterogeneity among the different sites, while, even if insulin/glucose was the most common therapy employed, in the majority of the patients multiple treatments were prescribed [37].

5. The "Old-Fashioned" Sodium Polystyrene Sulfonate

Among the different possibilities for treating hyperkalemia, recently, attention has been focused on the GI elimination of potassium, mainly because of the availability of new drugs, such as patiromer and sodium zirconium cyclosilicate.

Indeed, historically, the only options for promoting K^+ elimination by the GI have been limited to the "old" cation-exchanging resin, sodium polystyrene sulfonate (SPS), and its derivate calcium polystyrene sulfonate. SPS is a benzene, diethenyl-polymer with ethenylbenzene-sulfonated sodium salt, whose reactive sulfonic groups exchange preloaded sodium for K^+ along the GI lumen (mostly in the large intestine) [38]. It can be given orally or as an enema, and it is often given with sorbitol to prevent constipation. Theoretically, the exchanging capacity of SPS is 1 mmol of potassium per 1 g of resin, but its efficiency in vivo may be lower than expected because sodium release is only partial [39]. Moreover, the peak effect is seen 4–6 h after the administration, and this is the reason why SPS is not indicated as an emergency intervention for hyperkaliemia [40].

SPS was approved for treatment of acute hyperkalemia in 1958, but surprisingly, despite its common use, its safety and efficacy have been poorly studied in controlled studies [41].

In 2011, analyzing a recent retrospective cohort of 122 patients (38% with CKD), Kessler et al. documented a possible direct dose–response relationship between SPS (used at a dose of 15, 30, 45 and 60 g) and a reduction in serum potassium [42]. Similarly, in 2015, in a randomized placebo-controlled trial involving 33 CKD patients with mild hyperkalemia, Lepage et al. found that SPS was superior to the placebo in reducing serum potassium over 7 days in patients with mild hyperkalemia and CKD [43]. So, these data confirm the clinical practice of using SPS as a part of the treatment of hyperkalemia.

The adverse effects of SPS include electrolyte disturbances, such as hypokalemia and hypomagnesemia, and gastrointestinal symptoms, such as nausea, constipation and diarrhea. However, severely adverse gastrointestinal effects, including ulceration, bleeding, ischemic colitis and perforation have also been reported, especially when combined with sorbitol [44]. In a retrospective case-control study including 123,391 inpatients, of whom 2194 were prescribed SPS, there was a doubling of the incidence of colonic necrosis between SPS users and non-users, which was not significant (0.14% vs. 0.07%, $P = 0.2$) [45]. Interestingly, a large retrospective population-based study from Canada similarly documented a significant two-fold increase in the incidences of hospitalization for serious adverse GI events in 20,020 SPS users, when compared with matched non-users [46]. Moreover, Laureati et al. examined SPS use and GI safety in a cohort including 3690 adults with CKD stages 4–5 (1288 on chronic dialysis) naive to SPS. They found that SPS initiation was associated with a higher incidence of severe GI adverse events, mainly ulcers and perforations, in a probable dose-dependent manner [47]. Beyond the GI side effects, it should be underlined that significative drug interactions have also been described while using SPS, and this could be relevant for cardiac and renal patients, who often take multiple pharmacological therapies [48].

Evaluating all these potential detriments associated with the chronic use of SPS, currently, the Food and Drug Administration (FDA) recommends the avoidance of SPS prescription for patients with active GI diseases or with a history of recent bowel surgery, and in any case the avoidance of taking SPS at the same time as any other oral medications [39]. These important limitations, together with the scarce data on SPS efficacy, may explain why there has been a need to develop new drugs for treating hyperkalemia by increasing GI potassium elimination.

6. Hyperkalemia: New Treatment Options

6.1. Patiromer

Patiromer FOS (for oral suspension), formerly known as RLY5016, was approved by the FDA in the USA in 2015, for the treatment of hyperkalemia.

Patiromer is a cross-linked polymer of 2-fluoro acrylic acid (91%), with divinylbenzenes (8%) and 1,7-octadiene (1%). It is used in the form of its calcium salt (ratio 2:1) and with sorbitol (one molecule per two calcium ions or four fluoroacrylic acid units, corresponding to 4 g of sorbitol for each 8.4 g of patiromer); a combination called patiromer sorbitex calcium [49].

It appears as a dry powder for oral suspension, made of insoluble, spherical beads, with an average particle size of ≈ 100 μm. Pharmacokinetic analysis in animals showed that patiromer is not absorbed from the gut, is not metabolized, and is excreted in an unchanged form in the feces [50]. Patiromer works by binding the free potassium ions in the gastrointestinal tract, mainly in the distal colon lumen, and releasing calcium ions for exchange, thus lowering the amount of potassium available for absorption and increasing the amount that is excreted via the feces. The net effect is a reduction of the potassium levels in the blood serum. In CKD patients, it has been demonstrated that patiromer at a dose of 8.4 g twice a day lowered potassium levels within 7 hours of administration. These levels continue to decrease for at least 48 hours if treatment is continued, and remain stable for 24 hours after the administration of the last dose [51].

Because of its delayed onset of action (4–7 h), patiromer cannot be used as an emergency treatment for hyperkalemia [52].

6.2. Efficacy Data

Under in vitro conditions mimicking the pH and potassium content of the colon, patiromer binds 8.5–8.8 mmol of potassium per gram of polymer, which is a 1.5- to 2.5-fold improvement over the other polymers. In 33 healthy volunteers, 4.2, 8.4 and 16.8 g of patiromer, administered for 8 days three times a day, caused a dose-dependent increase in fecal potassium excretion (all $P < 0.02$ vs. placebo), with a corresponding dose-dependent reduction in urinary extraction [53].

A similar effect was also found in small cohorts of CKD and hemodialysis hyperkalemic patients, including those receiving RAASi [54]. However, the efficacy, safety and tolerability of patiromer were also tested in large clinical trials, which enrolled patients at high risk of hyperkalemia, such as patients with CHF, diabetes and CKD (see Table 1).

The first study exploring the efficacy and safety of patiromer in a large population was the "Evaluation of Patiromer in Heart Failure Patients" (PEARL HF) study, which was a 4-week, multicenter, double-blind, placebo-controlled study designed to evaluate the use of patiromer in the prevention of hyperkalemia. A total of 105 normokalemic patients (K^+ 4.3–5.1 mmol/L) with CHF and either i) a history of hyperkalemia resulting in the discontinuation of ACEi/ARB/MRA and/or beta-blockers, or ii) CKD (eGFR < 60 mL/min) treated with one or more CHF therapies (ACEIs/ARBs and beta-blockers), were randomized to undertake double-blind treatment with 30 g/day patiromer or a placebo for 4 weeks, in association with spironolactone (at the initial dose of 25 mg/day, increased to 50 mg/day on day 15 if K^+ was ≤ 5.1 mmol/L) [55].

The endpoints included the change in serum K^+, the proportion of patients with hyperkaliemia ($K^+ > 5.5$ mmol/L) and the proportion titrated to spironolactone 50 mg/day. At the end of treatment, compared with the placebo group, the group on patiromer showed significantly lowered serum K^+ levels (−0.45 mmol/L, $P < 0.001$), a lower incidence of hyperkaliemia, and a higher proportion of patients on spironolactone 50 mg/day. Interestingly, in patients with CKD ($n = 66$), the difference in K^+ levels between groups was −0.52 mmol/L ($P = 0.031$), and the incidence of hyperkaliemia was 6.7% for patiromer vs. 38.5% for the placebo. Adverse events were mainly gastrointestinal, and mild or moderate in severity.

Furthermore, the AMETHYST-DN study was a multicenter, open-label, dose-ranging, phase 2 trial that evaluated the efficacy of patiromer in the treatment of hyperkalemia in type 2 diabetic patients, with diabetic nephropathy and CKD and receiving RAAS inhibitors (ACEi and/or ARB for at least 28 days) [56]. The primary endpoint was potassium reduction, from baseline to week 4 or before the start of dose titration. The mean age was 66 years, 86% of patients had CKD stage 3–4, and 35% had CHF. Hyperkalemic patients at screening were immediately randomized into the treatment phase, while normokalemic patients were re-evaluated after the adjustment of antihypertensive therapy with the addition of losartan and/or spironolactone. Overall, 306 hyperkalemic patients (serum K^+ 5–6 mmol/L) were eligible, and were stratified by potassium level into the categories of mild (5–5.5 mmol/L) and moderate (5.5–6 mmol/L) hyperkalemia, before being randomized to receive patiromer at increasing dosages (4.2 g, 8.4 g or 12.6 g bid in mild hyperkalemia, and 8.4 g, 12.6 g and 16.8 g bid for moderate hyperkalemia). The dosage was titrated to achieve a target serum $K^+ \leq 5$ mmol/L.

Patiromer significantly reduced serum potassium levels from the baseline in all patients in a similar manner for the different doses, regardless of the initial potassium levels and independently of other comorbidities, such as CHF, advanced CKD or resistant hypertension. Moreover, in all patients the potassium lowering began $\approx$ 48 h after starting the patiromer, while target levels were reached early by patients with mild hyperkalemia. Interestingly, the reduction in serum K^+ was achieved at week 8 (end of treatment phase) and maintained up to 52 weeks in patients who continued the treatment, whereas after discontinuation, serum potassium levels significantly increased. Regarding the safety profile, hypomagnesemia (7.2%) was the most common side effect, while constipation (6.3%) was the most common gastrointestinal adverse event. Moreover, patiromer treatment was also evaluated in a phase 3 study with CKD patients. So, in the OPAL-HK study, 237 patients with CKD stage 3–4 and serum potassium level of 5.1–6.5 mmol/L, undergoing stable treatment with one or more RAASi, were divided into two groups: those with mild hyperkalemia (serum K^+ 5.1–5.5 mmol/L) that received patiromer 4.2 g bid, and those with moderate to severe hyperkalemia (5.5–6.5 mmol/L) that received 8.4 g bid [57]. Then the patiromer dosage was titrated to reach and maintain a potassium level of 3.8–5.1 mmol/L, and the patients were followed-up for 4 weeks. The authors found that patiromer significantly lowered potassium levels from baseline to week 4 in the whole study population, and for all prespecified subgroups (age < or > 65 years, presence/absence of diabetes or CHF, and maximal or submaximal dose of RAASi). Notably, at week 4, 76% of the overall population reached the target serum potassium level. After this first phase, the study proceeded with a randomized phase, in which patients with a baseline potassium level of 5.5–6.5 mmol/L and who achieved a target serum potassium level were randomized to continue the same dosage of patiromer or switch to placebo, and were followed-up for an additional 8 weeks. Furthermore, in this case, patiromer showed its efficacy, since, unlike the patiromer group, patients taking the placebo presented a significant increase in serum K^+ (median K^+ increase of +0.72 mmol/L). This difference was also observed across the prespecified subgroups of patients (regardless of age, gender, baseline K^+ levels, diabetes, CHF and maximal/not maximal RAASi dosage) [58]. Moreover, a post-hoc analysis showed that the patiromer K-lowering efficacy and safety profile in CKD patients was not compromised by diuretic therapy [59].

An interesting opportunity offered by the potassium-lowering effects of patiromer has been explored in the recent phase 2 randomized AMBER study, which evaluated whether the use of patiromer allows a more persistent use of spironolactone in patients with CKD (eGFR 25 to $\leq$ 45 mL/min) and resistant hypertension [60]. 295 patients were randomly assigned to receive either placebo or patiromer (8.4 g once daily), in addition to open-label spironolactone (starting at 25 mg once daily). At week 12, 98 (66%) of the 148 patients in the placebo group, and 126 (86%) of the 147 patients in the patiromer group, remained on spironolactone, suggesting that patiromer can enable more patients to continue treatment with spironolactone under conditions in which this drug may be beneficial.

Although many data have been reported on the efficacy and safety of treatments with patiromer, on the other hand, several clinical studies are ongoing concerning the evaluation of patiromer in specific clinical settings. This is the case for the DIAMOND study, a phase 3b placebo-controlled and randomized trial, the intent of which is to determine if the patiromer treatment of CHF subjects with hyperkalemia while receiving RAASi allows the continued use of RAASi medications. Interestingly, this study will consider primary "hard" endpoints, constituted by the time to the first occurrence of cardiovascular death or hospitalization. The completion date of this trial is estimated as the middle of 2022 [61].

Table 1. Main clinical trials evaluating use of patiromer for chronic hyperkalemia.

Study, Year	Study Population	N	Study Design (with Patiromer Dosage)	Follow-Up (Weeks)	Main Results
PEARL-HF 2012 [55]	CHF, CKD or previous hyperkalemia causing RAASi interruption plus indication to start spironolactone	105	Randomized and double blind: patiromer 15 g bid vs. placebo Spironolactone starting dose 25 mg, progressive dose titration	4	Mean K$^+$ reduction: −0.45 mmol/L patiromer vs. placebo ($P < 0.001$)
AMETHYST-DN 2015 [56]	Diabetes plus CKD (stage 3–4) receiving RAASi with known hyperkalemia or those who developed hyperkalemia during run-in phase	306	Randomized and open label. Patients on ACEi or ARB started on spironolactone (1) Mild HK (5.1–5.5 mmol/L): Patiromer 4.2–8.4–12.6 g bid (2) Moderate HK (5.6–5.9 mmol/L): Patiromer 8.4 g– 12.6 g–16.8 g bid	52	(1) Mild HK: K$^+$ reduction − 0.35 mmol/L for 4.2 g, −0.51 mmol/L for 8.4 g, −0.55 mmol/L for 12.6 g (2) Moderate HK: K$^+$ reduction −0.87 mmol/L for 8.4 g −0.97 mmol/L for 12.6 g −0.92 for 16.8 g
OPAL-HK 2015 [57]	CKD patients (stage 3–4) on RAASi	243	Initial treatment phase: (1) Mild HK (K$^+$ 5.1–5.5 mmol/L) Patiromer 4.2 g bid (2) Moderate HK (K$^+$ 5.6–5.9 mmol/L) Patiromer 8.4 g bid	4	Mean K$^+$ reduction: −1.01 mmol/L vs. basal values
		107	Randomized maintenance phase: Continue patiromer ($n = 55$) vs. placebo ($n = 52$)	8	K$^+$ increase: +0.72 mmol/L in placebo vs. 0 mmol/L in patiromer ($P < 0.001$)

Abbreviations: CHF, Chronic heart failure; CKD, Chronic Kidney Disease; HK, Hyperkalemia; RAASi, Renin-Angiotensin-Aldosterone system inhibitors; bid, twice a day.

6.3. Safety and Tolerability

Patiromer was generally well tolerated. Overall, treatment-related adverse effects reported in the clinical trials occurred in ≈ 20% of the patients included.

They include electrolyte disorders, such as hypomagnesemia and hypokalemia, and mild gastrointestinal symptoms, such as constipation (8%), diarrhea (5%), nausea and flatulence [52]. In the product labeling, hypomagnesemia and hypokalemia are reported as adverse reactions in 5.3% and 4.7% of the treated patients, respectively [62].

Monitoring of serum magnesium is recommended, considering supplementations for patients who develop hypomagnesemia while on patiromer.

No cases of intestinal necrosis have been reported, probably as a consequence of the optimized characteristic of patiromer (i.e., uniform spherical shape, defined polymer bead size, low swelling ratio), which may improve the GI tolerability of this drug [63].

However, the use of patiromer is discouraged in patients with severe constipation, bowel obstruction or impaction, including abnormal post-operative bowel motility disorders, because of the potential ineffectiveness and the possibility of worsening gastrointestinal conditions [62].

6.4. Dosage, Administration and Drug Interactions

Based on the above-mentioned large trials, patiromer is recommended at a starting dosage of 8.4 g once daily, administered orally, which can be increased by 8.4-g increments per week, titrated up to a maximum of 25.2 g once daily.

There are limited data on the use of patiromer for dialysis patients. As such, currently, no dose adjustment is advised. Patiromer presents as a powder that can be mixed with water, apple juice or cranberry juice. It should be mixed in an initial volume of 40 mL of water, then stirred and more water added to obtain the desired consistency. Then, the mixture should be taken within 1 hour of initial suspension, and its results are equally effective and well-tolerated when taken without food or with food [64].

Finally, it should be underlined that in vitro studies indicated the possibility that patiromer may interact with some medications. In particular, in studies of healthy volunteers, the use of patiromer decreased the systemic exposure of coadministered ciprofloxacin, levothyroxine and metformin [65]. For these reasons, the administration of other oral medications at least 3 hours before or 3 hours after patiromer is recommended.

7. Sodium Zirconium Cyclosilicate

Sodium zirconium cyclosilicate (SZC), formerly known as ZS-9, is an insoluble, inorganic, non-polymer zirconium silicate compound, comprising units of oxygen-linked zirconium and silicon atoms in the form of a microporous cubic lattice framework.

It works as a selective cation exchange agent, primarily releasing hydrogen and sodium and preferentially capturing potassium, thus increasing its fecal excretion [66].

Its selectivity for potassium, which is > 25 times greater than that for calcium and magnesium ions, is due to the size of the pore, which is similar in diameter to unhydrated potassium (approximately 3 Å). Because of its high selectivity for potassium, SZC may bind it throughout the entire GI tract, and may exert a rapid K-lowering effect. It has been estimated that one gram of SZC binds about 3 mmol of potassium, and its activity begins within 1 h of the consumption [67].

At this stage, there are no studies comparing the pharmacodynamics properties of SZC when administered with or without food. Clinical studies have demonstrated that SCZ was not systemically absorbed, and no differences in urine and blood concentration were detected between treated and untreated patients.

7.1. Efficacy Data

SZC has been primarily evaluated in four randomized trials (ZS-002, ZS-003, ZS-004 and ZS-004E) and one open-label long-term study (ZS-005) (Table 2).

The ZS-002 was a phase 2 study investigating the safety, tolerability, efficacy and pharmacodynamics of SZC in 90 patients with stage 3 CKD and hyperkalemia (K^+ 5.0–6.0 mmol/L), who were randomized to receive SZC 0.3, 3 or 10 g three times a day, or a placebo. SZC showed a dose-dependent effect, and potassium levels significantly declined in the first 48 hours of the patients taking SZC at the doses of 3 g and 10 g (P = 0.048 and P < 0.0001, respectively, versus placebo) [68].

Then, SZC was also investigated in larger phase 3 randomized trials, where it showed a significant superiority to the placebo in achieving and maintaining normal serum potassium levels [69].

In particular, in the ZS-004 (Hyperkalemia Randomized Intervention Multidose ZS-9 Maintenance, HARMONIZE) trial, SZC safety and efficacy was tested in 258 patients with hyperkaliemia (K^+ > 5.1 mmol/L), who initially received SZC 10 g three times a day for 48 h [70]. Then, those achieving normokalemia (N = 237) were randomized to receive SZC 5, 10 or 15 g once daily, or a placebo for the next 28 days (double-blind maintenance phase). Both initial and maintenance phases were characterized by a significant dose-dependent reduction of K^+ in all SZC groups, compared with the placebo, even in the prespecified subgroups (CHF, diabetes, CKD and patients on RAASi). Compared with all other study groups, during the maintenance phase, there was a higher incidence of generalized and peripheral oedema in the SZC 15 g group (14.3%).

Designed as an open-label extension of the HARMONIZE trial, the ZS-004E study investigated the safety and efficacy of SZC in patients with hyperkalemia who completed ZS-004, or who discontinued ZS-004 due to hypokalemia or hyperkaliemia in the maintenance phase and had a mean K^+ of 3.5–6.2 mmol/L [71]. For 11 months, 123 patients received additional open-label treatment with SZC 10 g a day as an initial dose, which was then then titrated to maintain K^+ 3.5–5.0 mmol/L. During the study period, a serum potassium value ≤ 5.1 mmol/L (primary endpoint) was achieved in 100% of THE patients, and K^+ ≤ 5.5 mmol/L in 88.3%.

The long-term efficacy and safety of SZC were also investigated in the ZS-005 trial, a phase 3, prospective, open-label, single-arm,12-month study, in which 751 outpatients with hyperkalemia (K^+ > 5.1 mmol/L) were enrolled [72]. No dietary restrictions or changes in RAASi therapy were required. The starting dosage was SZC 10 g thrice daily, for 24 to 72h (correction phase), then those who reached serum K^+ 3.5–5.0 mmol/L at any point during the correction entered the maintenance phase (starting dose of SZC 5 g once daily). Dose titration (up to a maximum of 15 g daily, down to a minimum of 5 g every other day) was allowed based on serum potassium measurements. During the correction phase, 99% of patients achieved K^+ 3.5–5.5 mmol/L, while the proportions of patients who achieved mean K^+ ≤ 5.1mmol/L and ≤ 5.5 mmol/L across the maintenance were 88% and 99%, respectively.

Interestingly, a post hoc analysis of ZS-005 focused on the study of the subgroups of patients with CKD. Furthermore, in this case, SZC use was associated with a significant reduction in serum K^+ levels in the long-term maintenance phase, in a similar manner even when patients were stratified via baseline-estimated glomerular filtration rate (i.e., eGFR < 30 or > 30 mL/min) [73].

The recent phase 3, randomized, double-blind HARMONIZE-Global trial examined the efficacy and safety of SZC among outpatients with hyperkalemia, from diverse geographic and ethnic origins [74].

A total of 248 patients achieving normokalaemia following a 48-h correction phase, with thrice-daily SZC 10 g, were randomized to once-daily SZC 5 g, SZC 10 g or placebo during a 28-day maintenance phase. Both initial and maintenance SZC regimens were associated with a significant reduction in K^+ levels when compared to baseline values and placebo, and this effect lasted over the 28 days of treatment.

Table 2. Main clinical trials evaluating the use of sodium zirconium cyclosilicate for acute and chronic hyperkalemia.

Study, Year	Study Population	N	Study Design (with SCZ Dosage)	Follow-Up (Weeks)	Main Results
HARMONIZE 2014 [70]	K$^+$ > 5.1 mmol/L 69% CKD	258	Open-label: SZC 10 g tid No control group	48 h	Normokalemia (K$^+$ 3.5–5 mmol/L): 84% at 24 h 98% at 48 h
		237	Randomized normokalemic pts: Placebo vs. SZC 5 g–10 g–15 g single dose	4	Normokalemia: Placebo 46% SZC 5 g: 80%, SZC 10 g: 90% SZC 15 g: 94% (p < 0.001 vs. placebo for all SZC)
ZS-005 2019 [72]	K$^+$ ≥ 5.1 mmol/L 74% CKD	751	Open-label: SZC 10 g tid	72 h	Normokalemia (K$^+$ 3.5–5 mmol/L): 99%
		746	Maintenance phase SZC 5 g daily, titrated to 5g–15g No control group	52	K$^+$ < 5.1 mmol/L: Overall 88% months 3–12 (466 pts completed the trial)
ENERGIZE 2020 [75]	K$^+$ ≥ 5.8 mmol/L in ED	7	Randomized: SZC 10 g (n = 38) vs. Placebo (n = 32) up to 3 times in 10 h + (glucose-insulin)	1	Mean change in K$^+$ at 4 h: −0.41 mmol/L SZC −0.27 mmol/L placebo Mean change in K$^+$ at 2 h: −0.72 mmol/L SZC −0.36 mmol/L placebo
DIALIZE 2020 [75]	HD patients with predialysis K$^+$ ≥ 5.4 mmol/L		Randomized: SZC 5–15 g single dose in non-dialysis days vs. Placebo	8	K$^+$ 4–5 mmol/L: SZC group 41.2% placebo group 1.0%

Abbreviations: SZC, Sodium Zirconium Cyclosilicate; CHF, Chronic heart failure; CKD, Chronic Kidney Disease; HK, Hyperkalemia; RAASi, Renin-Angiotensin-Aldosterone system inhibitors; ED, Emergency Department; HD, hemodialysis; tid, three times a day.

However, besides general studies on the treatment of chronic hyperkalemia, SZC has also been tested in acute and specific clinical settings. So, based on the pharmacokinetics data and the findings of clinical trials that reported a rapid effect of SZC in lowering serum K^+, the authors of the ENERGIZE study explored the use of SZC in the ED [75]. It was a phase 2, multicenter, randomized, double-blind, placebo-controlled study, in which 70 patients with serum $K^+ > 5.8$ mmol/L admitted at the ED were randomized 1:1 to SZC 10 g or placebo, administered up to three times during a 10-h period, in association with insulin and glucose. Reductions in K^+ levels at 1 hour with SZC or the placebo were similar, probably due to the predominant potassium-lowering effect of the concomitant insulin and glucose treatment. A greater reduction in mean K^+ from the baseline was observed in the SZC group, compared with the placebo at 2 hours (-0.72 vs. -0.36 mmol/L, respectively), suggesting that SZC may provide an incremental benefit in the emergency treatment of hyperkalemia.

However, the K^+ level's reduction was not significantly different between the SZC and placebo groups when they were evaluated 4 h after drug consumption.

Instead, the authors of the DIALIZE study tested the capacity of SZC to reduce blood potassium levels among patients undergoing HD [76].

So, they performed a phase 3b, double-blind, randomized trial, in which 197 patients on maintenance HD and predialysis hyperkalemia were randomized to receive a placebo or SZC 5 g once daily, on non-dialysis day, and were titrated to maintaining normokalemia over 4 weeks in increments of 5 g, up to a maximum of 15 g. The primary efficacy outcome involved the proportion of patients maintaining pre-dialysis serum K^+ levels of 4–5 mmol/L, on three out of four dialysis treatments, after long interdialytic and not receiving rescue treatment. At the end of the study, 40 patients of the 97 receiving SZC (41.2%) met the primary endpoint, compared with 1 patient out of the 99 on placebo (1%). Interestingly, adverse effects, including interdialytic weight gain, were similar between the two groups. Thus, these findings suggest that SZC is an effective and well-tolerated treatment for predialysis hyperkalemia in HD patients (Table 2).

Further information on specific patient populations will be expected from the results of the ongoing PRIORITIZE-HF trial, which will evaluate SZC vs. placebo in patients with CHF taking RAASi. The completion date of this trial is estimated as the end of 2020 [77].

7.2. Safety and Tolerability

SZC is generally well tolerated. Hypokalemia occurred in 5.8% of patients enrolled in the ZS-005 trial [72].

In phase 2 and 3 trials, the incidence of gastrointestinal adverse events (nausea, constipation, vomiting or diarrhea) was similar between the treated group and the placebo group [78]. However, as was the case for patiromer, SZC should also not be used in patients with severe constipation, bowel obstructions or impaction, including abnormal postoperative bowel motility disorders [79].

A dose-related mild to moderate edema was observed in the SZC during the maintenance period (mostly in patients receiving maximum SZC dosage), but it was resolved spontaneously or with diuretic therapy. So, it is recommended to monitor signs of edema, especially in patients at risk of fluid overload, such CKD and CHF patients, probably adjusting dietary salt intake and the dose of diuretics [79].

Finally, a non-clinically relevant QTc interval prolongation, without an increased rate of arrhythmia, has been reported in some cases, probably as a consequence of the rapid decrease in serum potassium levels [69].

7.3. Dosage, Administration and Drug Interactions

The recommended starting dosage of SZC is 10 g three times a day; then, once normokalaemia is achieved (usually in 24–48 hours), the maintenance dosage is 5 g daily (the dosage can be titrated up to a maximum of 10 grams once daily, or down to a minimum of 5 g every other day).

For patients on dialysis, SZC should only be given on non-dialysis days at starting doses of 5 g once daily, followed by titrating the dose according to the pre-dialysis serum potassium value after the long inter dialytic interval [79].

SZC presents as a powder, and before consumption, the entire content of a sachet should be mixed with approximately 45 mL of water, and stirred well. It can be taken with or without food [66]. If hyperkaliemia persists after 72 hours with the maximum dosage, other treatment approaches should be considered.

SZC can transiently increase gastric pH, potentially affecting the absorption of co-administered drugs that exhibit pH-dependent solubility.

In vivo studies in healthy volunteers showed that, when co-administered with SZC, there was an increase in systemic exposure to weak acids, such as furosemide and atorvastatin, and a decrease in systemic exposures to weak bases, such as dabigatran [80].

So, the general advice is that other oral medications should be administered at least 2 hours before or 2 hours after SZC.

8. Conclusions

For decades, the absence of a therapeutic alternative to SPS has represented one of the main limitations to the management of hyperkalemia, especially in patients at high risk, such as those with CHF, diabetes and CKD undergoing treatment with RAASi.

Therefore, the development of new potassium-lowering agents, such as patiromer and SZC, has offered new opportunities for improving the management of hyperkalemia, even considering that, unlike SPS, these medications have proven their efficacy in large clinical trials in different clinical settings (see Table 3). Remarkably, patiromer and SZC appear to be well tolerated and safer compared to SPS, with the report of only mild GI disorders and no cases of intestinal necrosis.

However, although the available data are encouraging and support the use of patiromer and SZC in the management of hyperkalemia, several important issues remain to be explored [81].

For example, there are no data on compliance with the treatment, and no study has yet directly compared the efficacy and tolerability of patiromer with SZC.

Moreover, one of the main barriers to the use of the new potassium-lowering agents may be constituted by the higher cost of these treatments compared to SPS. There is thus a need to perform accurate cost-effectiveness analyses, also to evaluate the economic effects of the implementation of these new treatments. These analyses should consider the potential benefits derived from the reduced incidence of adverse effects, and from the optimization of chronic RAASi treatment, which, in turn, may improve clinical outcomes for CHF and CKD patients.

In this regard, it has been demonstrated by mathematical models that hyperkalemia prevention and treatment with patiromer is a potentially cost-effective intervention for the long-term maintenance of RAASi in patients at risk of hyperkalemia [82].

So, several studies are ongoing, and others should be designed to define the potentiality offered by the application of these new potassium binders in specific clinical settings, and to elucidate their roles in improving long-term clinical outcomes.

Table 3. Main characteristics of the approved potassium binders for the treatment of hyperkalemia.

Drug, FDA Approval	Mechanisms Location	Onset of Action	Patient Groups Tested in Clinical Trials	Adverse Effects	Cost
SPS, 1958	Non-specific organic ion-exchange resin. It exchanges sodium for Potassium. Colon	Variable, hours to days [39]	CKD, HD	Mild to moderate gastrointestinal effects, including colonic necrosis, poor tolerability, electrolyte disorders	Low
Patiromer, 2015	Non-specific organic ion-exchange resin. It exchanges calcium for potassium. Colon	Within 7 h [51]	CHF, Diabetes, CKD +/- mRAASi	Mild gastrointestinal effects, hypomagnesaemia, hypokalemia (3–6%)	Very high
SZC, 2018	Selective inorganic non-polymer. It exchanges sodium and hydrogen for potassium. Entire gastrointestinal tract	Median time 2 h [69]	CHF, CKD, HD ED +/- RAASi	Mild gastrointestinal effects, oedema and hypokalemia (dose-dependent)	Very high

Abbreviations: SPS, Sodium Polystyrene Sulfonate; SZC, Sodium Zirconium Cyclosilicate; CHF, Chronic heart failure; CKD, Chronic Kidney Disease; HK, Hyperkalemia; RAASi, Renin-Angiotensin-Aldosterone system inhibitors; ED, Emergency Department; HD, hemodialysis.

Author Contributions: P.E., N.E.C., M.S., V.F.; critical revision of the manuscript: L.C., D.P., F.C.; supervision and approval of the final draft: F.V. and G.G. All authors have read and agreed to the published version of the manuscript.

Funding: This research received no external funding.

Conflicts of Interest: The authors declare no conflict of interest.

References

1. Gumz, M.L.; Rabinowitz, L.; Wingo, C.S. An Integrated View of Potassium Homeostasis. *N. Engl. J. Med.* **2015**, *373*, 60–72. [CrossRef] [PubMed]
2. Kovesdy, C.P.; Appel, L.J.; Grams, M.E.; Gutekunst, L.; McCullough, P.A.; Palmer, B.F.; Pitt, B.; Sica, D.A.; Townsend, R.R. Potassium homeostasis in health and disease: Ascientific workshop cosponsored by the National Kidney Foundation and the American Society of Hypertension. *J. Am. Soc. Hypertens.* **2017**, *11*, 783–800. [CrossRef] [PubMed]
3. McDonough, A.A.; Youn, J.H. Potassium Homeostasis: The Knowns, the Unknowns, and the Health Benefits. *Physiology* **2017**, *32*, 100–111. [CrossRef] [PubMed]
4. Aggarwal, S.; Topaloglu, H.; Kumar, S. Trends in emergency room visits due to hyperkalemia in the United States. *Value Health* **2015**, *18*, A386. [CrossRef]
5. Horne, L.; Ashfaq, A.; MacLachlan, S.; Sinsakul, M.; Qin, L.; LoCasale, R.; Wetmore, J.B. Epidemiology and health outcomes associated with hyperkalemia in a primary care setting in England. *BMC Nephrol.* **2019**, *20*, 85. [CrossRef] [PubMed]
6. Acker, C.G.; Johnson, J.P.; Palevsky, P.M.; Greenberg, A. Hyperkalemia in hospitalized patients: Causes, adequacy of treatment, and results of an attempt to improve physician compliance with published therapy guidelines. *Arch. Intern. Med.* **1998**, *158*, 917–924. [CrossRef]
7. Jarman, P.R.; Kehely, A.M.; Mather, H.M. Hyperkalaemia in diabetes: Prevalence and associations. *Postgrad. Med. J.* **1995**, *71*, 551–552. [CrossRef]
8. Kovesdy, C.P.; Matsushita, K.; Sang, Y.; Brunskill, N.J.; Carrero, J.J.; Chodick, G.; Hasegawa, T.; Heerspink, H.L.; Hirayama, A.; Landman, G.W.; et al. Serum potassium and adverse outcomes across the range of kidney function: A CKD Prognosis Consortium meta-analysis. *Eur. Heart J.* **2018**, *39*, 1535–1542. [CrossRef]
9. Collins, A.J.; Pitt, B.; Reaven, N.; Funk, S.; McGaughey, K.; Wilson, D.; Bushinsky, D.A. Association of Serum Potassium with All-Cause Mortality in Patients with and without Heart Failure, Chronic Kidney Disease, and/or Diabetes. *Am. J. Nephrol.* **2017**, *46*, 213–221. [CrossRef]
10. Kovesdy, C.P. Epidemiology of hyperkalemia: An update. *Kidney Int. Suppl.* **2016**, *6*, 3–6. [CrossRef]
11. Weiss, J.N.; Qu, Z.; Shivkumar, K. Electrophysiology of Hypokalemia and Hyperkalemia. *Circ. Arrhythm. Electrophysiol.* **2017**, *10*, e004667. [CrossRef] [PubMed]
12. Hoppe, L.K.; Muhlack, D.C.; Koenig, W.; Carr, P.R.; Brenner, H.; Schöttker, B. Association of Abnormal Serum Potassium Levels with Arrhythmias and Cardiovascular Mortality: A Systematic Review and Meta-Analysis of Observational Studies. *Cardiovasc. Drugs Ther.* **2018**, *32*, 197–212. [CrossRef]
13. Wilson, N.S.; Hudson, J.Q.; Cox, Z.; King, T.; Finch, C.K. Hyperkalemia-induced paralysis. *Pharmacotherapy* **2009**, *29*, 1270–1272. [CrossRef] [PubMed]
14. Wang, A.Y. Optimally managing hyperkalemia in patients with cardiorenal syndrome. *Nephrol. Dial. Transplant.* **2019**, *34*, iii36–iii44. [CrossRef] [PubMed]
15. Welling, P.A. Roles and regulation of renal K channels. *Annu. Rev. Physiol.* **2016**, *78*, 415–435. [CrossRef]
16. Briet, M.; Schiffrin, E.L. Aldosterone: Effects on the kidney and cardiovascular system. *Nat. Rev. Nephrol.* **2010**, *6*, 261–273. [CrossRef]
17. Choi, H.Y.; Ha, S.K. Potassium balances in maintenance hemodialysis. *Electrolyte Blood Press.* **2013**, *11*, 9–16. [CrossRef]
18. Oh, K.S.; Oh, Y.T.; Kim, S.W.; Kita, T.; Kang, I.; Youn, J.H. Gut sensing of dietary K^+ intake increases renal K^+ excretion. *Am. J. Physiol. Regul. Integr. Comp. Physiol.* **2011**, *301*, R421–R429. [CrossRef]
19. Aronson, P.S.; Giebisch, G. Effects of pH on potassium: New explanations for old observations. *J. Am. Soc. Nephrol.* **2011**, *22*, 1981–1989. [CrossRef]

20. Aburto, N.J.; Hanson, S.; Gutierrez, H.; Hooper, L.; Elliott, P.; Cappuccio, F.P. Effect of increased potassium intake on cardiovascular risk factors and disease: Systematic review and meta-analyses. *BMJ* **2013**, *346*, f1378. [CrossRef]

21. Ellison, D.H.; Terker, A.S.; Gamba, G. Potassium and Its Discontents: New Insight, New Treatments. *J. Am. Soc. Nephrol.* **2016**, *27*, 981–989. [CrossRef]

22. Rabelink, T.J.; Koomans, H.A.; Hené, R.J.; Dorhout Mees, E.J. Early and late adjustment to potassium loading in humans. *Kidney Int.* **1990**, *38*, 942–947. [CrossRef] [PubMed]

23. Hunter, R.W.; Bailey, M.A. Hyperkalemia: Pathophysiology, risk factors and consequences. *Nephrol. Dial. Transplant.* **2019**, *34*, iii2–iii11. [CrossRef] [PubMed]

24. Sousa, A.G.; Cabral, J.V.; El-Feghaly, W.B.; de Sousa, L.S.; Nunes, A.B. Hyporeninemic hypoaldosteronism and diabetes mellitus: Pathophysiology assumptions, clinical aspects and implications for management. *World J. Diabetes* **2016**, *7*, 101–111. [CrossRef]

25. Cooper, L.B.; Lippmann, S.J.; Greiner, M.A.; Sharma, A.; Kelly, J.P.; Fonarow, G.C.; Yancy, C.W.; Heidenreich, P.A.; Hernandez, A.F. Use of Mineralocorticoid Receptor Antagonists in Patients with Heart Failure and Comorbid Diabetes Mellitus or Chronic Kidney Disease. *J. Am. Heart Assoc.* **2017**, *6*, e006540. [CrossRef] [PubMed]

26. Bandak, G.; Sang, Y.; Gasparini, A.; Chang, A.R.; Ballew, S.H.; Evans, M.; Arnlov, J.; Lund, L.H.; Inker, L.A.; Coresh, J.; et al. Hyperkalemia after initiating renin-angiotensin system blockade: The Stockholm creatinine measurements (SCREAM) project. *J. Am. Heart Assoc.* **2017**, *6*, 1–13. [CrossRef]

27. Fried, L.F.; Emanuele, N.; Zhang, J.H.; Brophy, M.; Conner, T.A.; Duckworth, W.; Leehey, D.J.; McCullough, P.A.; O'Connor, T.; Palevsky, P.M.; et al. Combined angiotensin inhibition for the treatment of diabetic nephropathy. *N. Engl. J. Med.* **2013**, *369*, 1892–1903. [CrossRef]

28. Ma, T.K.; Kam, K.K.; Yan, B.P.; Lam, Y.Y. Renin-angiotensin-aldosterone system blockade for cardiovascular diseases: Current status. *Br. J. Pharmacol.* **2010**, *160*, 1273–1292. [CrossRef]

29. Poggio, R.; Grancelli, H.O.; Miriuka, S.G. Understanding the risk of hyperkalaemia in heart failure: Role of aldosterone antagonism. *Postgrad. Med. J.* **2010**, *86*, 136–142. [CrossRef]

30. An, J.N.; Lee, J.P.; Jeon, H.J.; Kim, D.H.; Oh, Y.K.; Kim, Y.S.; Lim, C.S. Severe hyperkalemia requiring hospitalization: Predictors of mortality. *Crit. Care* **2012**, *16*, R225. [CrossRef]

31. Formiga, F.; Chivite, D.; Corbella, X.; Conde-Martel, A.; Arévalo-Lorido, J.C.; Trullàs, J.C.; Silvestre, J.P.; García, S.C.; Manzano, L.; Montero-Pérez-Barquero, M. Influence of potassium levels on one-year outcomes in elderly patients with acute heart failure. *Eur. J. Intern. Med.* **2019**, *60*, 24–30. [CrossRef] [PubMed]

32. Brueske, B.; Sidhu, M.S.; Schulman-Marcus, J.; Kashani, K.B.; Barsness, G.W.; Jentzer, J.C. Hyperkalemia Is Associated with Increased Mortality Among Unselected Cardiac Intensive Care Unit Patients. *J. Am. Heart Assoc.* **2019**, *8*, e011814. [CrossRef] [PubMed]

33. Dépret, F.; Peacock, W.F.; Liu, K.D.; Rafique, Z.; Rossignol, P.; Legrand, M. Management of hyperkalemia in the acutely ill patient. *Ann. Intensive Care* **2019**, *9*, 32. [CrossRef] [PubMed]

34. Liu, M.; Rafique, Z. Acute Management of Hyperkalemia. *Curr. Heart Fail. Rep.* **2019**, *16*, 67–74. [CrossRef]

35. Coutrot, M.; Dépret, F.; Legrand, M. Tailoring treatment of hyperkalemia. *Nephrol. Dial. Transplant.* **2019**, *34*, iii62–iii68. [CrossRef]

36. Batterink, J.; Cessford, T.A.; Taylor, R.A.I. Pharmacological interventions for the acute management of hyperkalaemia in adults. *Cochrane Database Syst. Rev.* **2015**, *10*, CD010344. [CrossRef]

37. Peacock, W.F.; Rafique, Z.; Clark, C.L.; Singer, A.J.; Turner, S.; Miller, J.; Char, D.; Lagina, A.; Smith, L.M.; Blomkalns, A.L.; et al. Real World Evidence for Treatment of Hyperkalemia in the Emergency Department (REVEAL-ED): A Multicenter, Prospective, Observational Study. *J. Emerg. Med.* **2018**, *55*, 741–750. [CrossRef]

38. Scherr, L.; Ogden, D.A.; Mead, A.W.; Spritz, N.; Rubin, A.L. Management of hyperkalemia with a cation-exchange resin. *N. Engl. J. Med.* **1961**, *264*, 115–119. [CrossRef]

39. Sanofi-Aventis U.S. LLC. Kayexalate® (Sodium Polystyrene Sulfonate [USP] Cation-Exchange Resin). Available online: http://www.accessdata.fda.gov/drugsatfda_docs/label/2009/011287s021lbl.pdf (accessed on 27 May 2020).

40. Parks, M.; Grady, D. Sodium Polystyrene Sulfonate for Hyperkalemia. *JAMA Intern. Med.* **2019**, *179*, 1023–1024. [CrossRef]

41. Labriola, L.; Jadoul, M. Sodium polystyrene sulfonate: Still news after 60 years on the market. *Nephrol. Dial. Transplant.* **2020**, gfaa004. [CrossRef]

42. Kessler, C.; Ng, J.; Valdez, K.; Xie, H.; Geiger, B. The use of sodium polystyrene sulfonate in the inpatient management of hyperkalemia. *J. Hosp. Med.* **2011**, *6*, 136–140. [CrossRef] [PubMed]

43. Lepage, L.; Dufour, A.C.; Doiron, J.; Handfield, K.; Desforges, K.; Bell, R.; Vallée, M.; Savoie, M.; Perreault, S.; Laurin, L.P.; et al. Randomized Clinical Trial of Sodium Polystyrene Sulfonate for the Treatment of Mild Hyperkalemia in CKD. *Clin. J. Am. Soc. Nephrol.* **2015**, *10*, 2136–2142. [CrossRef] [PubMed]

44. Thomas, A.; James, B.R.; Landsberg, D. Colonic necrosis due to oral kayexalate in a critically-ill patient. *Am. J. Med. Sci.* **2009**, *337*, 305–306. [CrossRef] [PubMed]

45. Watson, M.A.; Baker, T.P.; Nguyen, A.; Sebastianelli, M.E.; Stewart, H.L.; Oliver, D.K.; Abbott, K.C.; Yuan, C.M. Association of prescription of oral sodium polystyrene sulfonate with sorbitol in an inpatient setting with colonic necrosis: A retrospective cohort study. *Am. J. Kidney Dis.* **2012**, *60*, 409–416. [CrossRef] [PubMed]

46. Noel, J.A.; Bota, S.E.; Petrcich, W.; Garg, A.X.; Carrero, J.J.; Harel, Z.; Tangri, N.; Clark, E.G.; Komenda, P.; Sood, M.M. Risk of Hospitalization for Serious Adverse Gastrointestinal Events Associated with Sodium Polystyrene Sulfonate Use in Patients of Advanced Age. *JAMA Intern. Med.* **2019**, *179*, 1025–1033. [CrossRef] [PubMed]

47. Laureati, P.; Xu, Y.; Trevisan, M.; Schalin, L.; Mariani, I.; Bellocco, R.; Sood, M.M.; Barany, P.; Sjölander, A.; Evans, M.; et al. Initiation of sodium polystyrene sulphonate and the risk of gastrointestinal adverse events in advanced chronic kidney disease: A nationwide study. *Nephrol. Dial. Transplant.* **2019**, gfz150. [CrossRef]

48. Aschenbrenner, D.S. Potential Drug Interactions with Sodium Polystyrene Sulfonate. *Am. J. Nurs.* **2018**, *118*, 47. [CrossRef]

49. Li, L.; Harrison, S.D.; Cope, M.J.; Park, C.; Lee, L.; Salaymeh, F.; Madsen, D.; Benton, W.W.; Berman, L.; Buysse, J. Mechanism of action and pharmacology of patiromer, a nonabsorbed cross linked polymer that lowers serum potassium concentration in patients with hyperkalemia. *J. Cardiovasc. Pharmacol. Ther.* **2016**, *21*, 456–465. [CrossRef]

50. Blair, H.A. Patiromer: A review in hyperkalemia. *Clin. Drug Investig.* **2018**, *38*, 785–794. [CrossRef]

51. Bushinsky, D.A.; Williams, G.H.; Pitt, B.; Weir, M.R.; Freeman, M.W.; Garza, D.; Stasiv, Y.; Li, E.; Berman, L.; Bakris, G.L. Patiromer induces rapid and sustained potassium lowering in patients with chronic kidney disease and hyperkalemia. *Kidney Int.* **2015**, *88*, 1427–1433. [CrossRef]

52. Montaperto, A.G.; Gandhi, M.A.; Gashlin, L.Z.; Symoniak, M.R. Patiromer: A clinical review. *Curr. Med. Res. Opin.* **2016**, *32*, 155–164. [CrossRef] [PubMed]

53. Bushinsky, D.A.; Spiegel, D.M.; Gross, C.; Benton, W.W.; Fogli, J.; Gallant, K.M.; Du Mond, C.; Block, G.A.; Weir, M.R.; Pitt, B. Effects of patiromer on urinary ion excretion in healthy adults. *Clin. J. Am. Soc. Nephrol.* **2016**, *11810*, 1769–1776. [CrossRef] [PubMed]

54. Bushinsky, D.A.; Rossignol, P.; Spiegel, D.M.; Benton, W.W.; Yuan, J.; Block, G.A.; Wilcox, C.S.; Agarwal, R. Patiromer decreases serum potassium and phosphate levels in patients on hemodialysis. *Am. J. Nephrol.* **2016**, *44*, 404–410. [CrossRef] [PubMed]

55. Pitt, B.; Anker, S.D.; Bushinsky, D.A.; Kitzman, D.W.; Zannad, F.; Huang, I.Z. Evaluation of the efficacy and safety of RLY5016, a polymeric potassium binder, in a double-blind, placebo-controlled study in patients with chronic heart failure (the PEARL-HF) trial. *Eur. Heart J.* **2011**, *32*, 820–828. [CrossRef]

56. Bakris, G.L.; Pitt, B.; Weir, M.R.; Freeman, M.W.; Mayo, M.R.; Garza, D.; Stasiv, Y.; Zawadzki, R.; Berman, L.; Bushinsky, D.A. Effect of patiromer on serum potassium level in patients with hyperkalemia and diabetic kidney disease: The AMETHYST_DN randomized clinical trial. *JAMA* **2015**, *314*, 151–161. [CrossRef]

57. Weir, M.R.; Bakris, G.L.; Bushinsky, D.A.; Mayo, M.R.; Garza, D.; Stasiv, Y.; Wittes, J.; Christ-Schmidt, H.; Berman, L.; Pitt, B. Patiromer in patients with kidney disease and hyperkalemia receiving RAAS inhibitors. *N. Engl. J. Med.* **2015**, *372*, 211–221. [CrossRef]

58. Weir, M.R.; Bushinsky, D.A.; Benton, W.W.; Woods, S.D.; Mayo, M.R.; Arthur, S.P.; Pitt, B.; Bakris, G.L. Effect of Patiromer on Hyperkalemia Recurrence in Older Chronic Kidney Disease Patients Taking RAAS Inhibitors. *Am. J. Med.* **2018**, *131*, 555–564.e3. [CrossRef]

59. Weir, M.R.; Mayo, M.R.; Garza, D.; Arthur, S.A.; Berman, L.; Bushinsky, D.; Wilson, D.J.; Epstein, M. Effectiveness of patiromer in the treatment of hyperkalemia in chronic kidney disease patients with hypertension on diuretics. *J. Hypertens.* **2017**, *35* (Suppl. 1), S57–S63. [CrossRef]

60. Agarwal, R.; Rossignol, P.; Romero, A.; Garza, D.; Mayo, M.R.; Warren, S.; Ma, J.; White, W.B.; Williams, B. Patiromer versus placebo to enable spironolactone use in patients with resistant hypertension and chronic kidney disease (AMBER): A phase 2, randomised, double-blind, placebo-controlled trial. *Lancet* **2019**, *394*, 1540–1550. [CrossRef]

61. Available online: https://clinicaltrials.gov/ct2/show/NCT03888066 (accessed on 27 May 2020).

62. Available online: https://www.accessdata.fda.gov/drugsatfda_docs/label/2018/205739s016lbl.pdf (accessed on 27 May 2020).

63. Pitt, B.; Garza, D. The tolerability and safety profile of patiromer: A novel polymer-based potassium binder for the treatment of hyperkalemia. *Expert Opin. Drug Saf.* **2018**, *17*, 525–535. [CrossRef]

64. Pergola, P.E.; Spiegel, D.M.; Warren, S.; Yuan, J.; Weir, M.R. Patiromer Lowers Serum Potassium When Taken without Food: Comparison to Dosing with Food from an Open-Label, Randomized, Parallel Group Hyperkalemia Study. *Am. J. Nephrol.* **2017**, *46*, 323–332. [CrossRef]

65. Lesko, L.J.; Offman, E.; Brew, C.T.; Garza, D.; Benton, W.; Mayo, M.R.; Romero, A.; Du Mond, C.; Weir, M.R. Evaluation of the Potential for Drug Interactions with Patiromer in Healthy Volunteers. *J. Cardiovasc. Pharmacol. Ther.* **2017**, *22*, 434–446. [CrossRef] [PubMed]

66. Stavros, F.; Yang, A.; Leon, A.; Nuttall, M.; Rasmussen, H.S. Characterization of structure and function of ZS-9, a K+ selective ion trap. *PLoS ONE* **2014**, *9*, e114686. [CrossRef] [PubMed]

67. Available online: https://ec.europa.eu/health/documents/community-register/2018/20180322137333/anx_137333_en.pdf (accessed on 27 May 2020).

68. Ash, S.R.; Singh, B.; Lavin, P.T.; Stavros, F.; Rasmussen, H.S. A phase 2 study on the treatment of hyperkalemia in patients with chronic kidney disease suggests that the selective potassium trap, ZS-9, is safe and efficient. *Kidney Int.* **2015**, *88*, 404–411. [CrossRef] [PubMed]

69. Packham, D.K.; Rasmussen, H.S.; Lavin, P.T.; El-Shahawy, M.A.; Roger, S.D.; Block, G.; Qunibi, W.; Pergola, P.; Singh, B. Sodium zirconium cyclosilicate in hyperkalemia. *N. Engl. J. Med.* **2015**, *372*, 222–231. [CrossRef] [PubMed]

70. Kosiborod, M.; Rasmussen, H.S.; Lavin, P.; Qunibi, W.Y.; Spinowitz, B.; Packham, D.; Roger, S.D.; Yang, A.; Lerma, E.; Singh, B. Effect of sodium zirconium cyclosilicate on potassium lowering for 28 days among outpatients with hyperkalemia: The HARMONIZE randomized clinical trial. *JAMA* **2014**, *312*, 2223–2233. [CrossRef]

71. Roger, S.D.; Spinowitz, B.S.; Lerma, E.V.; Singh, B.; Packham, D.K.; Al-Shurbaji, A.; Kosiborod, M. Efficacy and Safety of Sodium Zirconium Cyclosilicate for Treatment of Hyperkalemia: An 11-Month Open-Label Extension of HARMONIZE. *Am. J. Nephrol.* **2019**, *50*, 473–480. [CrossRef]

72. Spinowitz, B.S.; Fishbane, S.; Pergola, P.E.; Roger, S.D.; Lerma, E.V.; Butler, J.; von Haehling, S.; Adler, S.H.; Zhao, J.; Singh, B.; et al. Sodium Zirconium Cyclosilicate among Individuals with Hyperkalemia: A 12-Month Phase 3 Study. ZS-005 Study Investigators. *Clin. J. Am. Soc. Nephrol.* **2019**, *14*, 798–809. [CrossRef]

73. Roger, S.D.; Lavin, P.T.; Lerma, E.V.; McCullough, P.A.; Butler, J.; Spinowitz, B.S.; von Haehling, S.; Kosiborod, M.; Zhao, J.; Fishbane, S.; et al. Long-term safety and efficacy of sodium zirconium cyclosilicate for hyperkalaemia in patients with mild/moderate versus severe/end-stage chronic kidney disease: Comparative results from an open-label, Phase 3 study. *Nephrol. Dial. Transplant.* **2020**, gfz285. [CrossRef]

74. Zannad, F.; Hsu, B.G.; Maeda, Y.; Shin, S.K.; Vishneva, E.M.; Rensfeldt, M.; Eklund, S.; Zhao, J. Efficacy and safety of sodium zirconium cyclosilicate for hyperkalaemia: The randomized, placebo-controlled HARMONIZE-Global study. *ESC Heart Fail.* **2020**, *7*, 54–64. [CrossRef]

75. Peacock, W.F.; Rafique, Z.; Vishnevskiy, K.; Michelson, E.; Vishneva, E.; Zvereva, T.; Nahra, R.; Li, D.; Miller, J. Emergency Potassium Normalization Treatment Including Sodium Zirconium Cyclosilicate: A Phase II, Randomized, Double-blind, Placebo-controlled Study (ENERGIZE). *Acad. Emerg. Med.* **2020**, *27*, 475–486. [CrossRef]

76. Fishbane, S.; Ford, M.; Fukagawa, M.; McCafferty, K.; Rastogi, A.; Spinowitz, B.; Staroselskiy, K.; Vishnevskiy, K.; Lisovskaja, V.; Al-Shurbaji, A.; et al. A Phase 3b, Randomized, Double-Blind, Placebo-Controlled Study of Sodium Zirconium Cyclosilicate for Reducing the Incidence of Predialysis Hyperkalemia. *J. Am. Soc. Nephrol.* **2019**, *30*, 1723–1733. [CrossRef]

77. Available online: https://clinicaltrials.gov/ct2/show/NCT03532009?cond=Sodium+Zirconium+Cyclosilicate&draw=2&rank=8 (accessed on 27 May 2020).

78. Hoy, S.M. Sodium Zirconium Cyclosilicate: A Review in Hyperkalaemia. *Drugs* **2018**, *78*, 1605–1613. [CrossRef] [PubMed]

79. Available online: https://www.accessdata.fda.gov/drugsatfda_docs/label/2018/207078s000lbl.pdf (accessed on 24 June 2020).

80. Levien, T.L.; Baker, D.E. Sodium Zirconium Cyclosilicate. *Hosp. Pharm.* **2019**, *54*, 12–19. [CrossRef] [PubMed]

81. Bianchi, S.; Regolisti, G. Pivotal clinical trials, meta-analyses and current guidelines in the treatment of hyperkalemia. *Nephrol. Dial. Transplant.* **2019**, *34*, iii51–iii61. [CrossRef]

82. Sutherland, C.S.; Braunhofer, P.G.; Vrouchou, P.; van Stiphout, J.; Messerli, M.; Suter, K.; Zadok, N.; Schwenkglenks, M.; Ademi, Z. A cost-utility analysis of RAASi-enabling patiromer in patients with hyperkalemia. *Value Health* **2017**, *20*, A490. [CrossRef]

Journal of
Clinical Medicine

Article

Lifestyle-Related Exposure to Cadmium and Lead Is Associated with Diabetic Kidney Disease

Ilse J. M. Hagedoorn [1,*], Christina M. Gant [2,3,*], Sanne v. Huizen [1], Ronald G. H. J. Maatman [4], Gerjan Navis [3], Stephan J. L. Bakker [3] and Gozewijn D. Laverman [1]

[1] Division of Nephrology, Department of Internal Medicine, Ziekenhuisgroep Twente, 7609 PP Almelo, The Netherlands; s.vanhuizen1@vumc.nl (S.v.H.); g.laverman@zgt.nl (G.D.L.)

[2] Department of Internal Medicine, Meander Medisch Centrum, 3813 TZ Amersfoort, The Netherlands

[3] Division of Nephrology, Department of Internal Medicine, University of Groningen, University Medical Center Groningen, 9713 GZ Groningen, The Netherlands; g.j.navis@umcg.nl (G.N.); s.j.l.bakker@umcg.nl (S.J.L.B.)

[4] Department of Clinical Chemistry, Medlon BV, 7512 KZ Enschede, The Netherlands; r.maatman@zgt.nl

* Correspondence: ilse_hagedoorn10@hotmail.com (I.J.M.H.); cm.gant@meandermc.nl (C.M.G.); Tel.: +31-88-7083418 (I.J.M.H.); Fax: +31-88-7087081 (I.J.M.H.)

Received: 11 June 2020; Accepted: 28 July 2020; Published: 30 July 2020

Abstract: Background: Environmental factors contributing to diabetic kidney disease are incompletely understood. We investigated whether blood cadmium and lead concentrations were associated with the prevalence of diabetic kidney disease, and to what extent lifestyle-related exposures (diet and smoking) contribute to blood cadmium and lead concentrations. Material and methods: In a cross-sectional analysis in 231 patients with type 2 diabetes included in the DIAbetes and LifEstyle Cohort Twente (DIALECT-1), blood cadmium and lead concentrations were determined using inductively coupled plasma mass spectrometry. The associations between diet (derived from food frequency questionnaire), smoking and cadmium and lead were determined using multivariate linear regression. The associations between cadmium and lead and diabetic kidney disease (albumin excretion >30 mg/24 h and/or creatinine clearance <60 mL/min/1.73 m^2) were determined using multivariate logistic regression. Results: Median blood concentrations were 2.94 nmol/L (interquartile range (IQR): 1.78–4.98 nmol/L) for cadmium and 0.07 µmol/L (IQR: 0.04–0.09 µmol/L) for lead, i.e., below acute toxicity values. Every doubling of lead concentration was associated with a 1.75 (95% confidence interval (CI): 1.11–2.74) times higher risk for albuminuria. In addition, both cadmium (odds ratio (OR) 1.50 95% CI: 1.02–2.21) and lead (OR 1.83 95% CI: 1.07–3.15) were associated with an increased risk for reduced creatinine clearance. Both passive smoking and active smoking were positively associated with cadmium concentration. Alcohol intake was positively associated with lead concentration. No positive associations were found between dietary intake and cadmium or lead. Conclusions: The association between cadmium and lead and the prevalence of diabetic kidney disease suggests cadmium and lead might contribute to the development of diabetic kidney disease. Exposure to cadmium and lead could be a so far underappreciated nephrotoxic mechanism of smoking and alcohol consumption.

Keywords: alcohol; cadmium; diabetic kidney disease; diet; lifestyle related exposures; lead; proteinuria; smoking; type 2 diabetes

---

1. Introduction

Diabetic kidney disease (DKD) is one of the most debilitating complications in patients with type 2 diabetes (T2D) [1]. Although effective treatment options are available for reduction of albuminuria, blood pressure control and glycemic regulation, progression of DKD into end-stage kidney disease

(ESKD) is still common, underlining the necessity to identify additional mechanisms to target for renoprotection [2–4]. Exposure to the heavy metals cadmium (Cd) and lead (Pb) could be interesting in this respect. Cd and Pb bind to low-molecular-weight proteins, which are freely filtered through the glomerulus and then reabsorbed by the proximal tubules, causing primary tubular toxicity [5]. This may lead to albuminuria and progressive kidney disease towards ESKD [6].

High-grade exposure to Cd and/or Pb is undoubtedly nephrotoxic [5–7]. Moreover, evidence suggests that low blood levels of Cd and/or Pb already have unwarranted effects, as associations have been found between limited Cd and/or Pb blood concentrations and renal tubular defect markers, reduced eGFR and/or albuminuria [8–13]. These associations appear to be dose-dependent [8–12].

Despite a general reduction of industrial exposure to Cd and Pb over the past decades, exposure to these metals is still present in the population at a lower grade, in particular, through smoking and ingestion of food (from contaminated soil and water) [7,14].

Patients with T2D are at risk of developing DKD and may be more vulnerable to the nephrotoxic effects of low-grade Cd and Pb exposure [8,15–19]. However, epidemiological studies on the nephrotoxic effects of low-level Cd and Pb exposure in patients with T2D are scarce. Therefore, in this study in patients with T2D, we examine (1) the association between several lifestyle-related exposures, such as smoking and dietary intake, and the Cd and Pb blood concentrations and (2) the association between low-level Cd and Pb exposure and the prevalence of DKD.

2. Materials and Methods

2.1. Patient Inclusion

This study was performed in the DIAbetes and LifEstyle Cohort Twente-1 (DIALECT-1), which was previously described in detail [20]. All adult patients with T2D treated in the outpatient clinic internal medicine/nephrology in the Ziekenhuisgroep Twente Hospital, Almelo and Hengelo, the Netherlands, were eligible for participation. Exclusion criteria were ESKD and inability to understand the informed consent procedure. Patients were included between 2009 and 2016. The study was performed in accordance with the Helsinki agreement and the guidelines of good clinical practice. Prior to participation, all patients signed an informed consent form. DIALECT was approved by the local institutional review boards (METC-registration numbers NL57219.044.16 and 1009.68020) and was registered in the Netherlands Trial Register (NTR trial code 5855).

2.2. Data Collection

Information on medical conditions and medication use was obtained from electronic patient files and verified with the patient during the baseline visit. Information on smoking habits was collected through questionnaires. Diet and alcohol consumption were assessed with a Food-Frequency Questionnaire, which was previously validated [21]. Anthropometric measurements and presence of diabetic polyneuropathy were obtained from physical examination. The body surface area (BSA) was calculated using the universally adopted formula of DuBois [22]. Blood pressure was measured in supine position with an automated device (Dinamap®; GE Medical systems, Milwaukee, WI, USA) for 15 min with one-minute intervals. The mean systolic and diastolic pressure of the last three measurements was used to estimate the mean arterial pressure (MAP), which was used for further analysis. The MAP was calculated by the following formula: (2× diastolic blood pressure + systolic blood pressure)/3. Microvascular complications were defined as the presence of polyneuropathy, diabetic kidney disease and/or retinopathy. Macrovascular complications were defined as the presence of peripheral arterial disease, coronary artery disease and/or cerebrovascular disease. Venous blood and 24 h urine samples were stored at −80 °C for later analysis.

2.3. Measures of Diabetic Kidney Disease

In our cohort study, we aimed for extensive and deep phenotyping. Therefore we included collection of 24 h urine, to provide objective data on nutritional intake, including sodium intake. Furthermore, determination of 24 h urinary creatinine excretion allows for calculation of the glomerular filtration rate with serum creatinine while correcting for muscle mass of the individual, in contrast to the estimated glomerular filtration rate (eGFR) formulas, which correct serum creatinine for average muscle mass on a population scale. This allows not only for estimation of individual renal function with higher precision, but also for analyses with creatinine clearance as a continuous variable, because it is well-known that imprecision of eGFR as an estimate of renal function at higher estimated GFR levels is even greater than it is in the lower range [23]. So, we used creatinine clearance calculated from 24 h urinary creatinine excretion and corrected for BSA as primary endpoint [24]. DKD was defined as creatinine clearance <60 mL/min/1.73 m^2 and/or the presence of albuminuria (24 h urinary albumin excretion >30 mg/day). Secondary analyses were performed with albuminuria based on albumin/creatinine ratio in 24 h-urine or morning void (>2.5 mg/mmol for men and >3.5 mg/mmol for women).

2.4. Cd and Pb Measurements

Blood Cd and Pb concentrations were determined from EDTA whole blood. Samples were diluted 30× using 0.2% v/v HNO$_3$ 0.05% v/v Triton 1% v/v Methanol and analysed by inductively coupled plasma mass spectrometry (ICP-MS) using a kinetic energy discrimination procedure on the Perkin Elmer Nexion300× ICP-MS. Instrumentation settings are depicted in Supplementary Table S1. Levels below the limit of detection were entered just below the limit value. For Cd, 4 of the 240 patients had values below the limit of detection (LOQ 1 nmol/L). For Pb, no patients had values below the limit of detection (LOQ 21 nmol/L). Reference values were used as described, Cd < 5.0 µg/L [25] and Pb < 50.0 µg/L [26].

2.5. Statistical Analyses

All statistical analyses were performed using SPSS statistics (IBM SPSS Statistics for Windows, Version 23.0, Armonk, New York, NY, USA). Normality of data was determined by visual inspection of histograms. Data were presented as mean ± standard deviation (normal distribution), as median and interquartile range (IQR 25th–75th percentile, skewed data), or in number and percentage (categorical data). Patients with missing data on Cd and/or Pb were excluded from the study. Cases with otherwise missing data were excluded from the respective analyses. We performed transformation of the concentrations of Cd and Pb according to logarithm with base 2, which allows for interpretation of ORs per doubling of concentrations of Cd and Pb. Univariate linear regression analyses were performed to identify potential confounders in associations of Cd and Pb concentrations with variables of interest.

We used multivariate linear regression analyses to evaluate the determinants of blood concentration of Cd and Pb. A *p*-value < 0.05 was considered statistically significant. To evaluate the association between dietary intake and Cd and Pb concentration, the average total caloric intake and g/day of different food groups were ranked in tertiles. The two highest tertiles of several food products were compared with the lowest tertile in multivariate linear regression analyses. Supplementary Table S2 shows the components of the different food groups, i.e., vegetables, potatoes, liver and kidney, rice, bread, fish, fruit and cacao. To determine the association between Cd, Pb and DKD, a multivariate logistic regression analyses was performed using albuminuria (24 h urinary albumin excretion >30 mg/24 h) and creatinine clearance <60 mL/min/1.73 m^2 as primary outcome variables. Potential confounders were based on previous literature and univariate correlations. Potential interaction of associations of Cd and Pb with albuminuria and creatinine clearance <60 mL/min/1.73 m^2 by age, sex, smoking and alcohol intake was evaluated by inclusion of product-terms for these respective variables with Cd and Pb in the logistic regression analyses. In secondary analyses, logistic regression analyses was performed by employing the albumin/creatinine ratio as the outcome variable.

3. Results

3.1. Patient Characteristics, Cd and Pb Concentrations

Cd and Pb were determined in the first 240 patients included in DIALECT-1. In total, 231 patients were included in the analysis, and patients were excluded from the study if 24 h-urine was not available ($n = 1$) or if they were determined to be type 1 diabetics ($n = 8$). The characteristics of the study population ($n = 231$) are shown in Table 1. Mean age was 64 ± 9 years and the majority were men (60%). The median duration of diabetes was 12 (6–20) years. In total, 105 patients (46%) had DKD, of which 47 patients (20%) had a creatinine clearance <60 mL/min/1.73 m^2 and 86 patients (37%) had albuminuria (>30 mg/24 h). Mean HbA1c was $7.2 \pm 3.1\%$ (55 ± 10 mmol/mol) and 140 patients (61%) were on insulin therapy. The number of micro- (67%) and macrovascular complications (43%) was high. Median blood concentrations were 2.94 nmol/L (1.78–4.98 nmol/L) for Cd and 0.07 µmol/L (0.04–0.09 µmol/L) for Pb. All values were considered as otherwise 'normal' levels for Cd and Pb.

Table 1. Baseline characteristics and univariate associations with Cd and Pb.

Characteristics	*n=*	Total Population	Cadmium St. B.	*p*-Value	Lead St. B.	*p*-Value
Age, years	231	64 ± 9	−0.05	0.46	0.09	0.18
Women, *n* (%)	231	93 (40)	0.19	0.004	−0.12	0.06
Body Mass Index, kg/m^2	230	33 ± 6	−0.06	0.38	−0.04	0.57
Body surface area, m^2	230	2.1 ± 0.2	−0.14	0.03	0.06	0.37
Systolic blood pressure, mmHg	230	141 ± 17	−0.06	0.4	0.07	0.26
Diastolic blood pressure, mmHg	230	76 ± 10	0.02	0.74	0.05	0.46
Pulse rate, beats/min	229	75 ± 13	0.15	0.03	0.02	0.72
Diabetes duration, years	231	12 (6–20)	−0.22	0.001	−0.09	0.2
Treatment with insulin, *n* (%)	231	140 (61)	−0.2	0.003	−0.12	0.08
HbA1c, mmol/mol (%)	230	55 ± 10 (7.2 ± 3.1)	−0.14	0.04	−0.16	0.02
Smoking status	231					
Current active smokers, *n* (%)		35 (15)	0.46	<0.001	0.09	0.28
Current passive smokers, *n* (%)		46 (20)	0.15	0.05	0.04	0.63
Former active smokers, *n* (%)		95 (41)	0.15	0.05	0.18	0.03
Never smoked, *n* (%)		52 (23)	*		*	
Pack years, (years)	227	12 (0–29)	0.3	<0.001	0.23	<0.001
Alcohol, g/day	229	2.0 (0–14)	−0.1	0.15	0.3	<0.001
Creatinine clearance, mL/min/1.73 m^2	230	95 ± 39	−0.18	0.007	−0.12	0.08
eGFR, mL/min/1.73 m^2	231	75 ± 24	−0.12	0.08	−0.16	0.01
Urine albumin excretion, mg/24 h	223	11 (3–91)	0.12	0.08	0.13	0.05
Urine total protein excretion g/24 h	231	0.2 (0.1–0.4)	0.17	0.01	0.09	0.18
Blood cadmium, nmol/L	231	2.94 (1.78–4.98)			0.16	0.01
Blood lead, µmol/L	231	0.07 (0.04–0.09)	0.16	0.01		
Microvascular complications, *n* (%)	231	154 (67)	−0.03	0.67	0.19	0.003
Polyneuropathy, *n* (%)	231	78 (34)	−0.03	0.66	0.12	0.07
Diabetic kidney disease, *n* (%)	231	105 (46)	0.1	0.14	0.2	<0.001
Creatinine clearance <60 mL/min/1.73 m^2, *n* (%)	230	47 (20)	0.1	0.01	0.15	0.02
Albuminuria, *n* (%)	231	86 (37)	0.04	0.52	0.24	<0.001
Retinopathy, *n* (%)	231	64 (28)	−0.11	0.11	0.08	0.25
Macrovascular complications, n (%)	231	99 (43)	0.2	0.002	0.01	0.85
Peripheral arterial disease, *n* (%)	231	33 (14)	0.13	0.06	0.12	0.08
Coronary artery diseases, *n* (%)	231	66 (29)	0.09	0.18	0.01	0.9
Cerebrovascular diseases, *n* (%)	231	26 (11)	0.1	0.13	0.01	0.88

Standardized beta (St. B.) is shown. * never smoked is used as the reference value for smoking status. eGFR: estimated glomerular filtration rate.

3.2. Determinants of Cd and Pb Concentrations

Univariate associations between study parameters and Cd and Pb are shown in Table 1. Blood Cd concentration was higher in women ($\beta = 0.19$, $p = 0.004$) and in patients with exposure to smoking (pack years $\beta = 0.30$, $p < 0.001$; active smoking $\beta = 0.46$, $p < 0.001$). In addition, Cd was inversely associated with years of diabetes, insulin use and serum HbA1c. In multivariate analyses, the Cd concentration was significantly higher in active, passive and former smokers, compared to never smokers (Table 2). The association between alcohol and Cd remain statistically nonsignificant.

Table 2. Multivariate linear regression analyses on smoking status and alcohol intake and Cd and Pb concentrations.

Lifestyle-Related Exposures	Cadmium nmol/L		Lead umol/L	
	St. B.	*p*-Value	St. B.	*p*-Value
Active smoker				
Model 1 (crude)	0.46	<0.001	0.08	0.27
Model 2 (adjusted)	0.50	<0.001	0.11	0.16
Model 3 (adjusted)	0.48	<0.001	0.02	0.86
Passive smoker				
Model 1 (crude)	0.16	0.04	−0.003	0.97
Model 2 (adjusted)	0.17	0.03	−0.001	0.99
Model 3 (adjusted)	0.17	0.03	−0.03	0.69
Former smoker				
Model 1 (crude)	0.17	0.03	0.14	0.08
Model 2 (adjusted)	0.25	0.002	0.17	0.05
Model 3 (adjusted)	0.22	0.005	0.12	0.16
Alcohol intake				
Model 1 (crude)	−0.09	0.17	0.29	<0.001
Model 2 (adjusted)	0.003	0.96	0.30	<0.001
Model 3 (adjusted)	−0.05	0.49	0.30	<0.001

Standardized beta (St. B.) is shown. Alcohol intake in g/day. Never-smoked is used as the reference value for smoking status. Model 1 is unadjusted (crude), while Model 2 is adjusted for age, sex, creatinine clearance and Model 3 is adjusted for Model 2 and lead (for cadmium) and cadmium (for lead).

With respect to blood Pb concentration, both smoking (pack years $\beta = 0.23$, $p < 0.001$, former smoking $\beta = 0.18$ $p = 0.03$) and alcohol intake ($\beta = 0.30$, $p < 0.001$) were univariately associated with Pb concentrations. Alcohol intake remained significantly associated with Pb concentration in the fully adjusted model ($\beta = 0.30$, $p < 0.001$), while the association between smoking and Pb was not observed in the fully adjusted model (Table 2). Cd and Pb concentrations were associated with one another ($\beta = 0.16$, $p = 0.01$).

3.3. Dietary Intake and Blood Cd and Pb Concentrations

The average daily intake of different food products is shown in Supplementary Table S3. The mean caloric intake was 1850 ± 619 kcal/day. No positive associations were found between the two highest tertiles of different food products and the Cd and Pb concentrations, neither after adjustment for possible confounders and the total caloric intake (Table 3). Interestingly, a significant inverse association was observed between the highest tertiles of vegetables and fruits and the Cd concentration compared with the lowest tertiles ($\beta = -0.16$, $p = 0.03$ and $\beta = -0.14$, $p = 0.05$, respectively).

3.4. Multivariate Analysis between Cd and Pb and Diabetic Kidney Disease

Subsequently, we investigated the associations between Cd and Pb, and 24 h urinary albumin excretion (>30 mg/24 h) and creatinine clearance <60 mL/min/1.73 m^2 (Table 4). Doubling of the Pb concentration was strongly associated with albuminuria (OR 1.75, 95% CI: 1.11–2.74), which remained unchanged in the fully adjusted model. There was no significant association between Cd and albuminuria. Secondary analysis with albumin/creatinine ratio as the outcome variable instead of 24 h albuminuria yielded similar results (Supplementary Table S4). There was a significant association between Cd and albumin/creatinine ratio, however this association may be accounted for mostly by smoking and Pb as this association became statistically nonsignificant when adjusting for these confounders. Doubling of the Cd and Pb concentrations was significantly associated with creatinine clearance <60 mL/min/1.73 m^2 (respectively, OR 1.50 95% CI: 1.02–2.21 and 1.83 95% CI: 1.07–3.15), even when adjusting for the other metal (Table 4). We found no indication for effect-modification of associations of Cd and Pb with albuminuria and reduced creatinine clearance (all *p*-values for interaction terms >0.10).

Table 3. Multivariate linear regression between different food products as independent variables and the Cd and Pb concentrations as dependent variables.

Food Products	Cadmium (nmol/L)				Lead (µmol/L)			
	Crude		Adjusted *		Crude		Adjusted *	
	St. B.	*p* Value	St. B.	*p* Value	St. B.	*p* Value	St. B.	*p* Value
Vegetable								
Tertile 2	−0.15	0.06	−0.13	0.06	0.06	0.47	0.09	0.21
Tertile 3	−0.16	0.03	−0.16	0.03	−0.02	0.81	0.01	0.93
Rice								
Tertile 2	−0.01	0.95	−0.02	0.81	−0.08	0.32	−0.06	0.42
Tertile 3	−0.06	0.42	−0.02	0.81	−0.16	0.14	−0.1	0.2
Potatoes								
Tertile 2	−0.12	0.13	−0.12	0.11	0.14	0.08	0.11	0.14
Tertile 3	−0.13	0.1	−0.13	0.1	0.02	0.83	0.02	0.86
Bread								
Tertile 2	−0.07	0.34	−0.05	0.48	0.07	0.36	0.1	0.18
Tertile 3	−0.15	0.06	−0.06	0.46	0.02	0.83	0.08	0.31
Fish								
Tertile 2	0.01	0.9	−0.01	0.9	0.05	0.53	0.07	0.32
Tertile 3	−0.02	0.77	0.2	0.78	0.05	0.56	0.02	0.84
Fruit								
Tertile 2	−0.18	0.02	−0.08	0.24	−0.03	0.66	−0.02	0.8
Tertile 3	−0.15	0.05	−0.14	0.05	0.03	0.69	0	0.99
Liver and kidney								
Tertile 2	−0.12	0.09	−0.8	0.25	−0.15	0.03	−0.11	0.1
Tertile 3	−0.16	0.02	−0.12	0.07	−0.01	0.88	0.02	0.8
Cacao								
Tertile 2	−0.11	0.18	−0.07	0.32	−0.05	0.53	−0.01	0.1
Tertile 3	−0.08	0.3	−0.03	0.66	−0.1	0.22	−0.02	0.76

* Adjusted for: age, sex, total caloric intake, creatinine clearance (mL/min/1.73 m^2), pack years, alcohol intake (g/day), Pb (for Cd) and Cd (for Pb). Cd and Pb as dependent continuous variables.

Table 4. Multivariate logistic regression on the association between Cd and Pb and albuminuria and reduced creatinine clearance.

Independent Variables	Creatinine Clearance <60 mL/min/1.73 m^2		Albuminuria >30 mg/24 h	
	OR	95% CI	OR	95% CI
Cadmium nmol/L				
Model 1 (crude)	1.57	1.14–2.16	1.08	0.85–1.37
Model 2	1.65	1.18–2.31	1.22	0.95–1.58
Model 3	1.53	1.08–2.17	1.24	0.95–1.62
Model 4	1.52	1.07–2.16	1.24	0.95–1.63
Model 5	1.57	1.07–2.30	1.06	0.80–1.41
Model 6 *	1.50	1.02–2.21	1.01	0.75–1.36
Lead µmol/L				
Model 1 (crude)	1.65	1.07–2.55	2.15	1.44–3.19
Model 2	1.68	1.06–2.66	1.98	1.31–2.98
Model 3	1.63	1.01–2.66	1.97	1.29–3.01
Model 4	1.94	1.16–3.24	1.93	1.25–3.00
Model 5	1.94	1.15–3.27	1.75	1.12–2.74
Model 6 **	1.83	1.07–3.15	1.75	1.11–2.74

Odds ratio (OR) and 95% confidence interval (CI) are shown. Model 1 is unadjusted (crude), Model 2 is adjusted for age, sex, Model 3 is adjusted for Model 2 and HbA1c, insulin use, years diabetes, mean arterial pressure, Model 4 is adjusted for Model 3 and alcohol intake (g/day), Model 5 is adjusted for Model 4 and pack years and Model 6 is adjusted for Model 5 and * lead (for cadmium) and ** cadmium (for lead). The ORs for the additional independent variables of Model 6 are listing in Supplementary Table S5.

4. Discussion

Cadmium (Cd) and Lead (Pb) are two of the most prevalent and nephrotoxic heavy metals [27]. Evidence suggests that patients with T2D are more susceptible to renal toxic effects of Cd and Pb [8,15–19,28]. In our study, with concentrations of Cd and Pb considerably below the values for acute toxicity, we found clear associations between these elements and albuminuria and reduced creatinine clearance, respectively, in patients with T2D. Doubling of the Pb concentration was associated with a 1.75 times higher risk of albuminuria. Additionally, both higher Cd and Pb revealed an increased risk of creatinine clearance <60 mL/min/1.73 m^2. Furthermore, smoking and alcohol intake appear to be associated with the Cd and Pb concentrations, respectively.

Although no lower limit of blood Pb and Cd for (nephro)toxicity had been established previously [18,27], the Cd and Pb concentrations found in this study were all in the supposedly normal range, as found in Europe and the United States [8,9,11]. In the general population, previous studies have described the associations between Cd and Pb exposure and reduced eGFR or creatinine clearance, but with higher blood levels, i.e., Cd ≥0.60 μg/L (5.34 nmol/L) and Pb >1.82 μg/dL (0.09 μmol/L) [8–11]. The current study is the first to investigate these associations in a population of both men and women with complicated T2D [28]. We used two different, creatinine-based measures to assess renal function, i.e., eGFR and creatinine clearance. While the association with creatinine clearance was significant for Cd, and the associations with eGFR for Pb, with borderline associations between creatinine clearance and Pb, and between eGFR and Cd, respectively, the consistency of the direction of the associations supports the robustness of our findings.

With respect to albuminuria, we found no positive association between Cd and albuminuria, which is in line with previous studies in the general population [11,29]. In secondary analyses, the association between Cd and albumin/creatinine ratio may be attributed mostly to smoking (pack years) and Pb concentrations. However, we did find higher concentrations of Cd in women, and it has been shown previously that women with T2D have a higher risk for albuminuria from cadmium exposure compared to women without T2D [28]. Moreover, Madrigal and colleagues found that the association between Cd and eGFR was more pronounced among females in the general population [12]. When the interaction term between Cd and sex was added to the adjusted models for creatinine clearance <60 mL/min/1.73 m^2 and albuminuria, we found no significant interactions ($p = 0.54$, $p = 0.89$, respectively).

For Pb, several prospective studies in patients with and without T2D have found that environmental Pb accelerates progressive kidney disease (based on the creatinine clearance) [17,19,30,31], overruling negative findings in cross-sectional studies [9,11]. Two studies in patients with T2D have shown that Pb concentrations of 4–6 times higher than our population, were associated with an increased long-term risk for progressive kidney disease [17,19]. Moreover, the glomerular filtration rate improved in these patients after lead-chelation therapy [17]. Whether these Pb concentrations are 'low-level' is debatable, and due to sociodemographic differences, these results cannot be extended to our study population. We found a strong association between Pb and albuminuria and reduced creatinine clearance in a much lower concentration than previously described. Although causality cannot be proven, this suggests that Pb exposure may enhance DKD in a much lower concentration than previously thought. Possibly, patients with T2D are more susceptible to renal damage due to Pb exposure than the general population, or the mechanism of nephrotoxicity of Pb in T2D differs from the mechanism in the nondiabetic population.

Our study was not designed to investigate the mechanisms of renal damage by Cd and Pb, but in the literature several mechanisms have been implicated [5–7,15,19,27,32]. Even though both metals bind to low-molecular-weight proteins and primarily affect the proximal tubules, renal outcomes tend to diverge—while Cd-induced renal impairment is characterized in the early stages by the presence of increased excretion of LMWH proteins (β2-microglobuline, retinol binding protein and a1-microglobulin), in Pb nephropathy, the proteinuria (including albuminuria) is absent or minimal in the early stages of renal diseases [27].

However, both metals seem to interfere with diabetes metabolism in a way that might interact with the process of DKD. For Cd, the background of hyperglycaemia appears to provide an environment which promotes Cd-induced renal impairment in diabetes [15]. In diabetic obese mice, a Cd-induced proteinuria increase was achieved with a 4-fold lower Cd than those in nondiabetic control mice [33]. Moreover, Cd exposure causes abnormal adipocyte differentiation, expansion and function, which might lead to development of insulin resistance, hypertension and cardiovascular diseases [34]. We found that longer duration of diabetes and higher HbA1c were associated with lower cadmium levels, which is somewhat counterintuitive, but in line with findings of a previous study in patients with established T2D [35]. A possible explanation is that this is the consequence of healthy survivor bias. For Pb, one hypothesis is that low-level environmental Pb exposure can lead to oxidative stress reactions causing functional nitric oxide deficiency and activation of the renin-angiotensin-aldosterone system in patients with diabetes [19]. Lead chelation therapy can improve renal function through potential interference with this mechanism by reducing the reactive oxygen species [17,19].

The nephrotoxic effects of low-level Cd exposure in the general population are still under debate [11,29,36,37]. Several studies have described a pattern of increased blood Cd, and decreased urine Cd concentrations combined with a decreased eGFR, which might suggest reverse causality (increased blood Cd concentrations due to reduced renal clearance) [11,29]. Therefore, prospective studies on the association between Cd and eGFR trajectory are warranted.

In our study, current smokers have a 2.5 times higher Cd concentration compared with nonsmokers. Additionally, even former- and passive smoking were positively associated with the Cd concentration, which is in line with previous findings [38]. Noting the association between Cd and reduced creatinine clearance, kidneys should be added to the long list of organs negatively affected by smoking. Although dietary intake of cereals, vegetables or shellfish is reported to be the most important source of cadmium in the nonsmoking population, we found no positive association between diet and Cd concentration [7,18]. This could be due to the fact that patients in this study have a relatively low intake of food categories which contain high amounts of Cd, for example only 12 patients reach the recommended average daily intake of 250 g vegetables per day. Interestingly, we found a negative association between the two highest tertiles of vegetables and fruit intake and Cd concentrations. This may be accounted for by residual confounding whereby more vegetable and fruit intake may reflect a healthier lifestyle, or maybe vegetables and fruit stimulate the clearance of cadmium in the body. For blood Pb concentration, alcohol intake was the most important contributing parameter. This might be explained by intake of Pb-contaminated alcoholic liquors [39]. According to a study into Pb in alcohol beverages, draught beers sampled contained greater than 10 µg/L of Pb and 4% contained greater than 100 µg/L of Pb [39]. Consumption of beer containing 50 µg/L of lead could make a substantial contribution to blood Pb concentrations in man. Consumption of 1l/day of wine containing 150 µg/L of lead could also make a major contribution to blood lead concentrations [39].

One strength of our study is the fact that a population of patients with T2D in only one geographic area was studied, and therefore environmental differences were minimized. Furthermore, albuminuria and renal function were based on the 24 h urine collection, instead of a single portion of urine as in most earlier studies. We found a lower creatinine clearance to be significantly associated with higher cadmium and lower eGFR to be significant associated with higher lead, while creatinine clearance was not significantly associated with lead level and eGFR was not significantly associated with cadmium level. It should be noted that although the latter associations were not significant, this was due to the fact that the magnitude of the associations was slightly smaller than that of the significant associations rather than that they were absent or inverse. With a larger sample size, these associations would likely also have been significant and congruent with the significant associations of lower creatinine clearance with higher cadmium and lower eGFR with higher lead. A limitation of our study is the cross-sectional design, allowing only research of associations rather than causality. However, even when the associations we found were based on reverse causality, this might be of great importance because it may implicate toxic accumulation of Cd and Pb. In patients with T2D maintenance haemodialysis, higher Cd levels were associated with increased hazard ratio for all-cause mortality (hazard ratio (HR) = 2.34 (1.10–4.96)) [40].

Although the overall Pb exposure levels have diminished in recent decades due to elimination of Pb from gasoline, there is still an under-recognized but persistent occurrence of Cd and Pb exposure in urban populations [32]. Due to industrial emissions and household waste from batteries, cosmetics, and painting, the soil becomes polluted and Cd and Pb end up in water and food products. Possibly, due to urbanisation, exposure to Cd from food products will increase in the coming years [7]. Considering that even low doses of Cd and Pb are likely to have harmful effects, the lifestyle-related exposures to Cd and Pb remain an important research topic which has several policy implications for public health.

In conclusion, our study revealed that in patients with T2D lifestyle-related Pb exposure is associated with the prevalence DKD in much lower concentrations than previously described. Although prospective studies are required to confirm a causal relationship, patients with T2D may be at increased risk for the toxic effects of low-level Cd and Pb exposure. Our study provides additional affirmation that avoiding possible exposure to Cd and Pb through smoking and alcohol intake is important in T2D.

Supplementary Materials: The following are available online at http://www.mdpi.com/2077-0383/9/8/2432/s1, Table S1: Perkin Elmer Nexion300× ICP–MS instrumentation, Table S2: Overview of food items included in each food category, Table S3: Daily amounts of dietary intake, Table S4: Multivariate logistic regression between Cd and Pb and albumin/creatinine ratio. Table S5: Multivariate logistic regression on the association between Cd and Pb and albuminuria and reduced creatinine clearance.

Author Contributions: I.J.M.H. researched data and wrote the manuscript, C.M.G., S.J.L.B. and G.D.L. researched data and reviewed/edited the manuscript. G.N. and S.v.H. reviewed/edited the manuscript. R.G.H.J.M. determined the Cd and Pb concentrations and contributed to the research design and methods. G.D.L. is the principal investigator of DIALECT and guarantor. All authors have read and agreed to the published version of the manuscript.

Funding: This study was financially supported by an unrestricted grant from AstraZeneca.

Acknowledgments: The authors would like to thank all students who have participated in DIALECT-1, Ziekenhuisgroep Twente, for their contribution to patient inclusion. The authors would like to thank Milou Oosterwijk for her assistance with the analyses with the food frequency questionnaire and MEDLON B.V. and staff members of the laboratory for the cooperation and their contribution to the determinations.

Conflicts of Interest: The authors declare no conflict of interest.

Abbreviations

BSA	Body surface area
T2D	Type 2 diabetes
Cd	Cadmium
DIALECT	DIAbetes and LifEstyle Cohort Twente
DKD	Diabetic kidney disease
ESKD	End stage kidney disease
ICP-MS	Inductively coupled plasma mass spectrometry
MAP	Mean arterial pressure
Pb	Lead

References

1. Atkins, R.C.; Zimmet, P. Diabetic kidney disease: Act now or pay later. *J. Bras. Nefrol.* **2010**, *23*, 7–10.
2. Ruggenenti, P.; Fassi, A.; Llieva, A.P.; Bruno, S.; Iliev, I.P.; Brusegan, V.; Rubis, N.; Gherardi, G.; Arnoldi, F.; Ganeva, M.; et al. Preventing microalbuminuria in type 2 diabetes. *N. Engl. J. Med.* **2004**, *351*, 1941–1951. [CrossRef] [PubMed]
3. Remuzzi, G.; Macia, M.; Ruggenenti, P. Prevention and Treatment of Diabetic Renal Disease in Type 2 Diabetes: The BENEDICT Study. *J. Am. Soc. Nephrol.* **2006**, *17*, S90–S97. [CrossRef] [PubMed]

4. Zoungas, S.; De Galan, B.E.; Ninomiya, T.; Grobbee, D.; Hamet, P.; Heller, S.R.; MacMahon, S.; Marre, M.; Neal, B.; Patel, A.; et al. Combined Effects of Routine Blood Pressure Lowering and Intensive Glucose Control on Macrovascular and Microvascular Outcomes in Patients With Type 2 Diabetes. *Diabetes Care* **2009**, *32*, 2068–2074. [CrossRef]

5. Madden, E.F.; Fowler, B.A. Mechanisms of nephrotoxicity from metal combinations: A review. *Drug Chem. Toxicol.* **2000**, *23*, 1–12. [CrossRef]

6. Johri, N.; Jacquillet, G.; Unwin, R. Heavy metal poisoning: The effects of cadmium on the kidney. *Biometals* **2010**, *23*, 783–792. [CrossRef]

7. Jarup, L.; Bergulund, M. Health effects of cadmium exposure—A review of the literature and risk estimate. *Scand. J. Work Environ. Health* **1998**, *24*, 1–52.

8. Åkesson, A.; Lundh, T.; Vahter, M.; Bjellerup, P.; Lidfeldt, J.; Nerbrand, C.; Samsioe, G.; Strömberg, U.; Skerfving, S. Tubular and Glomerular Kidney Effects in Swedish Women with Low Environmental Cadmium Exposure. *Environ. Health Perspect.* **2005**, *113*, 1627–1631. [CrossRef]

9. Navas-Acien, A.; Tellez-Plaza, M.; Guallar, E.; Muntner, P.; Silbergeld, E.; Jaar, B.; Weaver, V.M. Blood Cadmium and Lead and Chronic Kidney Disease in US Adults: A Joint Analysis. *Am. J. Epidemiol.* **2009**, *170*, 1156–1164. [CrossRef]

10. Ferraro, P.M.; Costanzi, S.; Naticchia, A.; Sturniolo, A.; Gambaro, G. Low level exposure to cadmium increases the risk of chronic kidney disease: Analysis of the NHANES 1999–2006. *BMC Public Health* **2010**, *10*, 1–8. [CrossRef]

11. Buser, M.C.; Ingber, S.Z.; Raines, N.; Fowler, D.A.; Scinicariello, F. Urinary and blood cadmium and lead and kidney function: NHANES 2007–2012. *Int. J. Hyg. Environ. Health* **2016**, *219*, 261–267. [CrossRef]

12. Madrigal, J.M.; Ricardo, A.C.; Persky, V.; Turyk, M. Associations between blood cadmium concentration and kidney function in the U.S. population: Impact of sex, diabetes and hypertension. *Environ. Res.* **2019**, *169*, 180–188. [CrossRef]

13. Buchet, J. Renal effects of cadmium body burden of the general population. *Lancet* **1990**, *336*, 699–702. [CrossRef]

14. Satarug, S. Dietary Cadmium Intake and Its Effects on Kidneys. *Toxics* **2018**, *6*, 15. [CrossRef]

15. Edwards, J.R.; Prozialeck, W.C. Cadmium, diabetes and chronic kidney disease. *Toxicol. Appl. Pharmacol.* **2009**, *238*, 289–293. [CrossRef]

16. Kim, N.H.; Hyun, Y.Y.; Lee, K.-B.; Chang, Y.; Rhu, S.; Oh, K.-H.; Ahn, C. Environmental Heavy Metal Exposure and Chronic Kidney Disease in the General Population. *J. Korean Med. Sci.* **2015**, *30*, 272–277. [CrossRef]

17. Lin, J.-L.; Lin-Tan, D.-T.; Yu, C.-C.; Li, Y.; Huang, Y.-Y.; Li, K.-L. Environmental exposure to lead and progressive diabetic nephropathy in patients with type II diabetes. *Kidney Int.* **2006**, *69*, 2049–2056. [CrossRef]

18. Satarug, S.; Haswell-Elkins, M.R.; Moore, M.R. Safe levels of cadmium intake to prevent renal toxicity in human subjects. *Br. J. Nutr.* **2000**, *84*, 791–802. [CrossRef]

19. Huang, W.-H.; Lin, J.-L.; Lin-Tan, D.-T.; Hsu, C.-W.; Chen, K.-H.; Yen, T.-H. Environmental Lead Exposure Accelerates Progressive Diabetic Nephropathy in Type II Diabetic Patients. *BioMed Res. Int.* **2013**, *2013*, 742545. [CrossRef]

20. Gant, C.M.; Binnenmars, S.H.; Berg, E.V.D.; Bakker, S.J.; Navis, G.J.; Laverman, G.D. Integrated Assessment of Pharmacological and Nutritional Cardiovascular Risk Management: Blood Pressure Control in the DIAbetes and LifEstyle Cohort Twente (DIALECT). *Nutrients* **2017**, *9*, 709. [CrossRef]

21. Feunekes, G.I.; Van Staveren, W.A.; De Vries, J.H.; Burema, J.; Hautvast, J.G. Relative and biomarker-based validity of a food-frequency questionnaire estimating intake of fats and cholesterol. *Am. J. Clin. Nutr.* **1993**, *58*, 489–496. [CrossRef]

22. Du Bois, D.; Du Bois, E.F. A formula to estimate the approximate surface area if height and weight be known. 1916. *Nutrients* **1989**, *5*, 303–311.

23. Coresh, J.; Astor, B.C.; McQuillan, G.; Kusek, J.; Greene, T.; Van Lente, F.; Levey, A.S. Calibration and random variation of the serum creatinine assay as critical elements of using equations to estimate glomerular filtration rate. *Am. J. Kidney Dis.* **2002**, *39*, 920–929. [CrossRef]

24. Inker, L.A.; Perrone, R.D. Calculation of the Creatinine Clearance. 2017. Available online: https://www-uptodate-com.proxy-ub.rug.nl/contents/calculation-of-the-creatinineclearance?search=Calculation%20of%20the%20creatinine%20clearance&source=search_result&selectedTitle=4~{}150&usage_type=default&display_rank=4 (accessed on 4 October 2019).
25. Moreau, T.; Lellouch, J.; Juguet, B.; Festy, B.; Orssaud, G.; Claude, J.R. Blood Cadmium Levels in a General Male Population with Special Reference to Smoking. *Arch. Environ. Health Int. J.* **1983**, *38*, 163–167. [CrossRef]
26. Nader, R.; Horwath, A.R.; Wittwer, C.T. Chapter 42: Toxic Elements. In *Tietz Textbook of Clinical Chemistry and Molecular Diagnostics*, 6th ed.; Elsevier: St. Louis, MO, USA, 2017; pp. 907–910.
27. Gonick, H.C. Nephrotoxicity of cadmium & lead. *Indian J. Med. Res.* **2008**, *128*, 335–352.
28. Barregard, L.; Bergström, G.; Fagerberg, B. Cadmium, type 2 diabetes, and kidney damage in a cohort of middle-aged women. *Environ. Res.* **2014**, *135*, 311–316. [CrossRef]
29. Chaumont, A.; Nickmilder, M.; Dumont, X.; Lundh, T.; Skerfving, S.; Bernard, A. Associations between proteins and heavy metals in urine at low environmental exposures: Evidence of reverse causality. *Toxicol. Lett.* **2012**, *210*, 345–352. [CrossRef]
30. Lin, J.-L.; Lin-Tan, D.-T.; Li, Y.-J.; Chen, K.-H.; Huang, Y.-L. Low-level Environmental Exposure to Lead and Progressive Chronic Kidney Diseases. *Am. J. Med.* **2006**, *119*, 707.e1–707.e9. [CrossRef]
31. Lin, J.-L.; Lin-Tan, D.-T.; Hsu, K.-H.; Yu, C.-C. Environmental Lead Exposure and Progression of Chronic Renal Diseases in Patients without Diabetes. *N. Engl. J. Med.* **2003**, *348*, 277–286. [CrossRef]
32. Leff, T.; Stemmer, P.; Tyrrell, J.; Jog, R. Diabetes and Exposure to Environmental Lead (Pb). *Toxics* **2018**, *6*, 54. [CrossRef]
33. Jin, T.; Nordberg, G.F.; Sehlin, J.; Leffler, P. The susceptibility of spontaneously diabetic mice cadmium-metallothionein nephrotoxicity. *Toxicology* **1994**, *89*, 81–90. [CrossRef]
34. Kawakami, T.; Sugimoto, H.; Furuichi, R.; Kadota, Y. Cadmium reduces adipocyte size and expression levels of adiponectin and Peg1/Mest in adipose tissue. *Toxicology* **2010**, *267*, 20–62. [CrossRef]
35. Anetor, J.I.; Uche, C.Z.; Ayita, E.B.; Adedapo, S.K.; Adeleye, J.O.; Anetor, G.O.; Akinlade, S.K. Cadmium Level, Glycemic Control, and Indices of Renal Function in Treated Type II Diabetics: Implications for Polluted Environments. *Front. Public Health* **2016**, *4*, 114. [CrossRef] [PubMed]
36. Weaver, V.M.; Kim, N.-S.; Lee, B.-K.; Parsons, P.J.; Spector, J.; Fadrowski, J.; Jaar, B.G.; Steuerwald, A.J.; Todd, A.C.; Simon, D.; et al. Differences in urine cadmium associations with kidney outcomes based on serum creatinine and cystatin C. *Environ. Res.* **2011**, *111*, 1236–1242. [CrossRef] [PubMed]
37. Akerstrom, M.; Sallsten, G.; Lundh, T.; Barregard, L. Associations between Urinary Excretion of Cadmium and Proteins in a Nonsmoking Population: Renal Toxity or Normal Physiology? *Environ. Health Perspect.* **2012**, *121*, 187–191. [CrossRef]
38. Jung, S.Y.; Kim, S.; Lee, K.; Kim, J.-Y.; Bae, W.K.; Lee, K.; Han, J.-S.; Kim, S. Association between secondhand smoke exposure and blood lead and cadmium concentration in community dwelling women: The fifth Korea National Health and Nutrition Examination Survey (2010–2012). *BMJ Open* **2015**, *5*, e008218. [CrossRef]
39. Sherlock, J.C.; Pickford, C.J.; White, G.F. Lead in alcoholic beverages. *Food Addit. Contam.* **1986**, *3*, 347–354. [CrossRef]
40. Yen, T.-H.; Lin, J.-L.; Lin-Tan, D.-T.; Hsu, C.-W.; Chen, K.-H.; Hsu, H.-H.; Pirozzi, N.; Apponi, F.; Napoletano, A.M.; Luciani, R.; et al. Blood cadmium level's association with 18-month mortality in diabetic patients with maintenance haemodialysis. *Nephrol. Dial. Transpl.* **2010**, *26*, 998–1005. [CrossRef]

 Journal of
Clinical Medicine

Article

Relationship among Left Ventricular Hypertrophy, Cardiovascular Events, and Preferred Blood Pressure Measurement Timing in Hemodialysis Patients

Hiroaki Io [1,2,*], Junichiro Nakata [2], Hiroyuki Inoshita [1,2], Masanori Ishizaka [1,2], Yasuhiko Tomino [2] and Yusuke Suzuki [2]

1   Department of Nephrology, Juntendo University Nerima Hospital, Tokyo 177-8521, Japan; ino-hi@juntendo.ac.jp (H.I.); m-ishiza@juntendo.ac.jp (M.I.)
2   Department of Nephrology, Juntendo University Faculty of Medicine, Tokyo 113-8412, Japan; jnakata@juntendo.ac.jp (J.N.); yasu@mtnet.jp (Y.T.); yusuke@juntendo.ac.jp (Y.S.)
*   Correspondence: hiroaki@juntendo.ac.jp; Tel.: +81-3-5923-3111

Received: 28 September 2020; Accepted: 27 October 2020; Published: 30 October 2020

Abstract: This study aimed to identify the ideal timing and setting for measuring blood pressure (BP) and determine whether the left ventricular mass index (LVMI) is an independent risk factor associated with increased cardiovascular events in hemodialysis (HD) patients. BP and LVMI were measured at baseline and at 6 and 12 months after HD initiation. BP was monitored and recorded at nine different time points, including before and after HD over a one-week period (HDBP). The mean BP measurement was calculated as the weekly averaged BP (WABP). LVMI was significantly correlated with home BP, in-office BP, HDBP, and WABP. Receiver operating characteristic analysis indicated that the cutoff LVMI value for cardiovascular events was 156 g/m^2. LVMI and diabetes mellitus were significant influencing factors for cardiovascular events (hazards ratio (95% confidence interval): diabetes mellitus, 2.84 (1.17,7.45); LVMI > 156 g/m^2, 2.86 (1.22,6.99)). Pre-HDBP, post-HDBP, and WABP were independently associated with higher LVMI in the follow-up periods. Hemoglobin and human atrial natriuretic peptide (hANP) levels were associated with LVMI beyond 12 months after HD initiation. Treatment of hypertension, overhydration based on hANP, and anemia may reduce the progression of LVMI and help identify HD patients at high risk for cardiovascular events.

Keywords: left ventricular hypertrophy; left ventricular mass index; hemodialysis; blood pressure; cardiovascular events; risk factor

---

1. Introduction

Cardiovascular (CV) disease is a leading cause of morbidity and mortality in hemodialysis (HD) patients. The US Renal Data System 2014 Annual Data Report stated that, between 2011 and 2013, cardiac arrest was the primary cause of two-thirds of CV-related deaths in HD patients [1]. In this population of patients, both volume and pressure overload result in increased cardiac work. The improvements in the prevention or postponement of kidney failure in the United States are possibly due to interventions such as greater blood pressure control in the general population. The prevalence of end-stage kidney disease (ESKD) continues to increase and reached 746,557 cases in 2017 (vs. 727,912 in 2016), representing a 2.5% increase since 2016, a reflection of decreasing mortality rates in the ESKD population [2]. The adaptive response to this physiological situation leads to left ventricular hypertrophy (LVH). LVH is a common comorbid condition observed with chronic kidney disease (CKD) and is a significant predictor of increased CV events in dialysis patients [3]. At the initiation of dialysis therapy, the prevalence of LVH is high [4], possibly due to both delayed diagnosis and insufficient treatment of hypertension.

Echocardiography provides an accurate estimate of LV mass (LVM). The diagnosis of LVH using echocardiography is frequently seen in patients with ESKD [5]. In early CKD patients, the LVM index (LVMI), as estimated by echocardiography, has been shown to correlate with the risk of progression to dialysis [6]. Lower blood hemoglobin level and higher LVMI are associated with progression to dialysis and poorer outcomes [7].

The optimal timing and measurement technique of BP in patients with ESKD are yet to be established. Currently, there is no established method of BP measurement that more accurately predicts the development of elevated LVMI and LVH and thus increased risk for CV events [8]. A significant change in BP, through a decrease in LV end-diastolic pressure, might decrease the left atrial (LA) diameter, altering the assessment of the LVH calculation of LVMI on echocardiography [9]. Many studies have used pre-HD BP to determine optimal BP levels in HD patients [10].

Moriya et al. [11] reported that the weekly averaged BP (WABP) is a useful method for estimating the BP of HD patients and correlates well with the LVMI. In HD patients, however, the relationship between LVH and BP variability remains unclear [12]. After starting HD, systolic BP, atrial natriuretic peptide (ANP), and hemoglobin levels have been found to be predictive factors for LVMI [3,9], more so than diastolic BP. Matsumoto et al. [13] reported that an increase in hemoglobin level along with a reduction in ANP were associated with reductions in LVH. Intensive HD is associated with lower risks for cardiovascular death and hospitalization, especially for heart failure, relative to both conventional HD and peritoneal dialysis [14]. In Japan, the incidence of ESKD has increased in the last few decades [15]. Although studies have reported an association between the prevalence of LVH and CV risk factors in patients with CKD [16], the relationship between LVMI and CV events in HD patients requires further exploration. Thus, this study aimed to assess when and how BP should be monitored in relation to HD and whether LVMI is an independent risk factor for increased CV events in HD patients.

2. Materials and Methods

The study protocol was approved by the Ethics Review Committee of Juntendo University Faculty of Medicine, Tokyo, Japan, in 2007 (approval no. 207-036, dated 16 October 2007) and registered with the University Hospital Medical Information Network (UMIN000018312). The modified study design (prospective observational study) was approved in 2015 (approval no. 22-78; 2015026). It complied with the tenets of the 2000 Declaration of Helsinki.

This prospective observational study was conducted in the Department of Nephrology of the Juntendo University Hospital. The inclusion criteria for the study were adults aged >20 years who were newly indicated for HD. Patients who were unable to revisit the outpatient clinic 6 months after initiating HD and those with concomitant malignancy, alcoholism, or chronic inflammatory disease were excluded.

Demographic data were collected at baseline. We measured the patients' human ANP (hANP), hemoglobin, high-sensitivity C-reactive protein (hsCRP), albumin, homocysteine, iron, and transferrin saturation levels at baseline (day 0), and at 6 and 12 months after initiation of HD. Of the 418 HD patients, 192 provided consent. A total of 72 patients were excluded from the study (22 had malignancy, 2 had alcohol abuse, 31 had inflammation or infection, 14 could not come to the hospital 6 months later due to relocation or hospital transfer, and 3 withdrew their consent), and the remaining 120 patients who survived beyond 6 months were included in this study.

2.1. BP Measurement

We measured and recorded 9 BP measurements over a one-week period for each patient. Home BP (HBP) was measured in the mornings on treatment days 1 and 5 of each week. BP was measured just before HD on each dialysis day (days 1, 3, and 5) (pre-HDBP) and again after each dialysis (post-HDBP). Finally, BP was measured once during their clinic echocardiogram visit day (day 4 of each week) (visit BP(VBP)). The WABP was defined as the average of these 9 BP measurements (Figure 1). Nine BP measurements were performed at baseline and at 6 and 12 months.

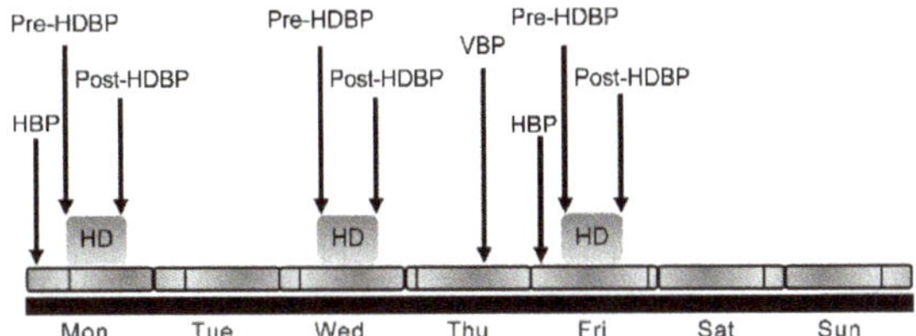

Figure 1. Blood pressure measurements over a one-week period. HBP, home blood pressure; VBP, visit blood pressure; HD; hemodialysis; HDBP, hemodialysis blood pressure.

2.2. Echocardiography Measurement

Echocardiographic examinations were conducted on a non-HD day. In all patients, echocardiographic examinations were performed by one examiner using the Toshiba ultrasound system model 260 SS-A equipped with a 2.5-MHz phased-array transducer (Toshiba, Tokyo, Japan). All examinations were performed with the patient lying in the left lateral recumbent position. Data were analyzed following the American Society of Echocardiography Guidelines. The LA and LV sizes, interventricular septal thickness, and LV posterior wall thickness were measured by two-dimensional and M-mode echocardiography [17]. We measured the LVMI [18], early/late LV filling velocity (E/A), and deceleration time. The LV mass was calculated using the following formula [19]:

$$\text{The LV mass} = 0.8\ (1.04\ (\text{LV internal diameter in diastole} + \text{posterior wall thickness} + \text{interventricular septal thickness})^3 - (\text{LV internal diameter in diastole})^3) + 0.6\ \text{g} \tag{1}$$

LVMI was defined as LV mass standardized by the body surface area [19].

2.3. CV Events

A CV event was defined as hospitalization for unstable angina, CV death, sustained arrhythmia, myocardial infarction, stroke, arteriosclerosis obliterans, transient ischemic attack, arterial aneurysm, and valve disease. Admission for fluid overload and pulmonary edema was included in those with a diagnosis of angina or myocardial infarction. Mere fluid overload due to insufficient dry weight settings was not included in the CV event.

2.4. Statistical Analysis

Statistical analysis was performed using JMP 10 (SAS Institute, Cary, NC, USA). Standard descriptive statistics were used to assess baseline characteristics. Data are presented as the mean ± standard deviation. Characteristics of patients with and without CV events were compared using analysis of variance (ANOVA). Variables with p values < 0.05 were further analyzed using a stepwise linear regression analysis based on a forward–backward procedure. Repeated-measures ANOVA was performed to compare serial changes in the clinical data and echocardiographic parameters. Cox regression was used to analyze relative CV event risks. We chose hANP as a measure of capacity load and BP as a measure of pressure load associated with LVH, in addition to LVMI, age, sex, diabetes mellitus (DM), BP, and LVMI, to evaluate the association between LVMI and CV events. The independence of these variables was evaluated to select variables to be used for multivariate Cox hazard analysis. A cutoff value of LVMI was calculated for CV events using receiver operating characteristic (ROC) analysis based on the value at 6–12 months after the start of dialysis, creating a binary categorical variable based on LVMI. The cumulative CV event incidence was assessed using the Kaplan–Meier method and log-rank test using significant influencing factors for CV events based on the multivariate Cox hazard model. Multivariate linear regression analysis was used to examine the association between LVMI or hemoglobin and clinical and laboratory variables. A value <0.05 was considered statistically significant.

3. Results

3.1. Patient Characteristics

A total of 418 patients who were admitted to our hospital between October 2008 and September 2012 for HD were screened. After applying the exclusion criteria, the final study population included 120 HD patients. The mean patient age was 63.5 ± 11.4 years at the initiation of HD. Most patients were male (73.3%), 42.6% had diabetes, and 30.8% had hypertensive nephrosclerosis. Diabetic kidney disease was defined as a patient with a history of diabetes for more than 5–10 years and diabetic retinopathy and/or high proteinuria.

3.2. Baseline and 12 Months Echocardiographic, Laboratory, and BP Values

Hemoglobin (g/dL), hANP (pg/mL), and hsCRP (mg/dL) levels differed significantly after 12 months of HD. In contrast, the LVMI at month 12 of HD (145.5 ± 46.2 g/m^2) was not significantly different from that at month 0 (178.1 ± 48.5 g/m^2). Only the end-of-week post-HD systolic BP decreased significantly after 12 months of HD (157.6 ± 18.5 vs. 136.8 ± 19.9 mmHg). Systolic and diastolic VBP, home systolic and diastolic BP (HBP), pre- and post-HD systolic BP, post-HD diastolic BP, and WABP did not change significantly at 12 months post-HD initiation.

3.3. Characteristics of Patients with and without CV Events 12 Months after Initiation of HD

Table 1 shows the differences between patients with and without CV events. There were no deaths. CV events included angina pectoris (n = 12), cerebral infarction (n = 6), occlusive arterial disease (n = 2), myocardial infarction (n = 1), and arrhythmia (n = 2). Age >70.6 years, DM, past CV event, and elevated hsCRP levels, were all significantly correlated with an elevated risk of CV events.

Table 1. ANOVA of characteristics between patients with and without CV events 12 months after initiation of HD.

	Patients with CV Events	Patients without CV Events	p Value	All Patients
Age (years)	70.6 ± 2.4	61.3 ± 1.4	<0.01	63.5 ± 11.4
Male sex (%)	77.3	72.5	0.65	73.6
DM (%)	63.6	36.1	0.02	42.6
Past CV event (%)	60.1	18.3	<0.01	27.7
Systolic blood pressure (mmHg)				
Outpatient clinic	135.2 ± 4.4	142.6 ± 2.3	0.15	140.9 ± 1.8
Before hemodialysis (start of the week)	148.2 ± 5.5	147.2 ± 3.1	0.87	147.4 ± 2.1
Mean blood pressure	143.1 ± 3.5	141.1 ± 3.5	0.63	141.6 ± 1.1
Laboratory parameters				
Hemoglobin (g/dL)	10.9 ± 0.3	10.6 ± 0.1	0.24	10.7 ± 0.1
Non-dialysis day hemoglobin (g/dL)	11.5 ± 0.4	11.5 ± 0.4	0.87	11.5 ± 0.2
Total iron saturation (%)	22.1 ± 3.5	27.9 ± 23.8	0.15	26.4 ± 14.7
Albumin (g/dL)	3.79 ± 0.07	3.77 ± 0.04	0.84	3.77 ± 0.03
Homocysteine (mg/dL)	23.7 ± 6.2	34.7 ± 3.1	0.11	32.7 ± 2.2
hsCRP (mg/dL)	0.49 ± 0.09	0.13 ± 0.05	<0.01	0.21 ± 0.04
hANP (pg/ dL)	68.4 ± 12.5	60.5 ± 6.3	0.58	62.1 ± 4.9
Alkaline phosphatase (U/L)	258.1 ± 18.7	232.1 ± 10.2	0.23	238.1 ± 8.2
Corrected calcium (mg/dL)	8.69 ± 0.28	8.97 ± 0.16	0.39	8.89 ± 0.12
Phosphorous (mg/dL)	5.31 ± 0.29	5.39 ± 0.16	0.74	5.38 ± 0.13
Total cholesterol (mg/dL)	158.3 ± 8.2	168.3 ± 4.7	0.29	165.8 ± 3.7
LDL cholesterol (mg/dL)	84.6 ± 7.9	90.9 ± 4.3	0.49	89.4 ± 3.4
Intact PTH (pg/dL)	171.1 ± 29.4	152.8 ± 15.4	0.58	156.7 ± 12.1
Erythropoietin stimulating agent (U/week)	21.1 ± 3.2	18.9 ± 1.9	0.54	19.4 ± 1.4
Echocardiographic data				
LAD (mm)	37.7 ± 1.3	35.8 ± 0.7	0.21	36.3 ± 0.6
Relative wall thickness	0.528 ± 0.02	0.496 ± 0.01	0.16	0.503 ± 0.01
LVMI (g/m^2)	155.4 ± 7.9	144.6 ± 4.2	0.23	146.9 ± 3.4
EF (%)	66.8 ± 2.3	68.2 ± 1.2	0.61	67.9 ± 0.9
Follow-up period (days)	532.4 ± 104.2	728.1 ± 57.2	0.11	682.8 ± 48.8

ANOVA, analysis of variance; CV, cardiovascular; DM, diabetes mellitus; HD, hemodialysis; hsCRP, high-sensitivity C-reactive protein; hANP, human atrial natriuretic peptide; LDL, low-density lipoprotein; PTH, parathyroid hormone; LAD, left atrial dimension; LVMI, left ventricular mass index; EF, ejection fraction.

3.4. Cox Proportional Hazards Modeling

As a result of the ROC analysis, the calculated cutoff value of LVMI for CV events was 156 g/m^2. LVMI, age, DM, sex, hANP, and BP (outpatient systolic BP at 6–12 months) were considered as candidates independent variables in the examination of the influence of LVMI on CV events. Evaluation of the independence of these variables showed that LVMI was significantly correlated with hANP and BP (outpatient systolic BP). In addition, hANP, BP, and age were all significantly different between the two LVMI groups. Since the number of events in this instance was 23, the independent variable that can be input to the Cox proportional hazards model was approximately 1/10 of the events, which was approximately 2 to 3. The current variable was selected because only approximately three independent variables were available. BP, hANP, and age were excluded from the above six variable candidates, and analysis using a multivariate Cox hazard model was performed using LVMI, sex, and DM as independent variables. As a result, the model showed significance (p = 0.0068): LVMI and DM were found to be significant influencing factors for CV events (hazards ratio (95% confidence interval): DM, 2.84 (1.17–7.45); LVMI >156 g/m^2, 2.86 (1.22–6.99)) (Table 2).

Table 2. Adjusted hazard ratio for CV events.

Variables	HR (95% CI)	p Value
Male sex	1.44 (0.45,6.41)	0.5595
DM	2.84 (1.17,7.45)	0.0214
LVMI > 156 g/m^2	2.86 (1.22,6.99)	0.0154

Multivariate Cox hazard model: p = 0.0068. CV, cardiovascular; CI, confidence interval; DM, diabetes mellitus; HR, hazard ratio; LVMI, left ventricular mass index.

A significant difference was found in the cumulative CV event incidence rates among the four subgroups classified according to the LVMI classification and the presence or absence of DM, both of which were significant influencing factors in the Cox hazard model (p = 0.0059) (Figure 2).

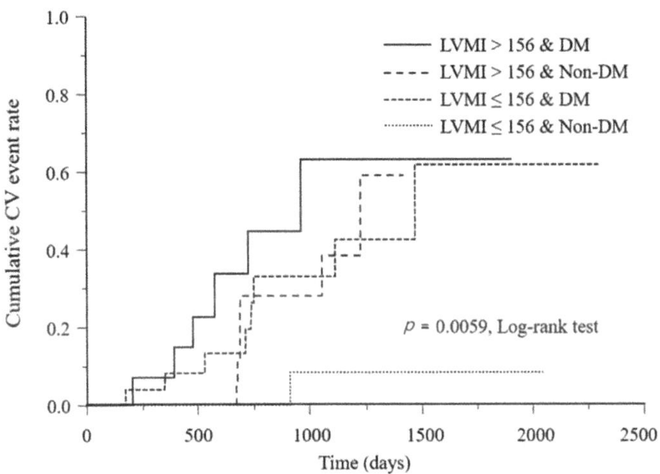

Figure 2. Kaplan–Meier analysis of CV event incidence according to LVMI class and DM. CV, cardiovascular; LVMI, left ventricular mass index; DM, diabetes mellitus.

Sex, BP, and hANP were the factors not related to CV events that were considered in the univariate analysis. All patients who were included in the analysis were taking angiotensin receptor blockers or angiotensin-converting enzyme inhibitors. Furthermore, the analysis of the cardiothoracic ratio (CTR), a simple method for evaluating dry weight in dialysis patients of Japanese origin, was not possible in our cohort because the CTR data included both pre- and post-dialysis values.

We presented the baseline characteristic with the patients grouped according to LVMI <156 g/m^2 and LVMI >156 g/m^2 (Table 3).

Table 3. Comparison of baseline characteristics between patients with LVMI < 156 g/m^2 and patients with LVMI > 156 g/m^2.

	LVMI < 156 g/m^2	LVMI > 156 g/m^2	p Value	All Patients
Age (years)	61.4 ± 13.2	63.7 ± 11.9	0.37	63.0 ± 11.4
Male sex (%)	67.5	73.7	0.51	71.8
DM (%)	41.2	46.8	0.58	45.1
Systolic blood pressure (mmHg)				
Outpatient clinic	139.8 ± 18.5	147.7 ± 18.2	<0.05	144.8 ± 18.4
Before hemodialysis (start of the week)	147.2 ± 26.5	147.6 ± 19.5	0.95	147.5 ± 22.1
Mean blood pressure	142.9 ± 18.9	140.5 ± 13.8	0.63	141.3 ± 15.1
Labolatory parameters				
Hemoglobin (g/dL)	9.1 ± 1.1	8.8 ± 1.1	0.36	8.9 ± 1.1
TSat (%)	21.7 ± 8.8	28.2 ± 16.4	0.11	26.2 ± 14.6
Albumin (g/dL)	3.75 ± 0.37	3.77 ± 0.28	0.84	3.77 ± 0.32
Homocysteine (mg/dL)	37.8 ± 28.8	31.6 ± 20.2	0.32	33.6 ± 22.2
hsCRP (mg/dL)	0.18 ± 0.08	0.24 ± 0.06	0.59	0.22 ± 0.06
hANP (pg/dL)	50.9 ± 29.6	65.4 ± 57.1	0.24	60.6 ± 44.9
Alkaline phosphatase (U/L)	256.2 ± 70.7	230.3 ± 85.1	0.18	239.4 ± 81.2
Phosphorous (mg/dL)	5.3 ± 0.8	5.3 ± 1.4	0.89	5.3 ± 1.3
Calcium-phosphorous product (mg^2/dL^2)	47.5 ± 8.3	48.3 ± 12.9	0.77	48.1 ± 11.2
Total cholesterol (mg/dL)	168.5 ± 41.1	160.9 ± 30.1	0.36	163.4 ± 35.7
LDL cholesterol (mg/dL)	91.1 ± 42.1	88.3 ± 27.1	0.73	89.2 ± 33.1
Intact PTH (pg/dL)	185.4 ± 23.9	137.6 ± 16.8	0.11	153.4 ± 18.1

LVMI, left ventricular mass index; DM, diabetes mellitus; TSat, transferrin saturation; hsCRP, high-sensitivity C-reactive protein; hANP, human atrial natriuretic peptide; LDL, low-density lipoprotein; PTH, parathyroid hormone.

3.5. Factors Associated with LVMI 12 Months after Initiation of HD

The univariate regression analysis showed that LVMI was significantly correlated with start- and end-of-week waking systolic HBP, systolic VBP, pre-HD systolic BP, post-HD systolic BP, WABP, and hemoglobin levels at 12 months after HD initiation (Table 4). hANP was independently associated with LVMI after 6 months and 12 months (Table 4). Hemoglobin levels were significantly correlated with end-of-week post-HD systolic BP. LVMI was positively correlated with systolic blood pressure at almost all blood pressure measurements. Sex, LVMI (binary variable classified by 156 g/m^2), and blood pressure were selected as independent variables, and blood pressure was evaluated in each model using VBP, WABP, start of week before HD-BP, and end of week after HD-BP. For significant models, the p value for each variable and the odds ratio [95% confidence interval] to past CV events were calculated. In model 4, high LVMI and end of week after HD-BP were found to be associated with past CV events. In other words, high LVMI showed an increase in the incidence of past CV events, and low blood pressure after weekend dialysis showed an increase in past CV events (Table 5).

Table 4. Univariate analysis of factors associated with LVMI beyond 12 months after HD initiation.

Factor	*r* Value	*p* Value
Hemoglobin (HD day)	−0.154	<0.001 *
Hemoglobin (non-HD day)	−0.118	0.08
hANP	0.514	<0.0001 *
hsCRP	0.112	0.083
Homocysteine	−0.164	0.019 *
Albumin	−0.176	0.0017 *
Iron	−0.111	0.04 *
TSat	−0.107	0.22
Systolic VBP	0.287	<0.0001 *
Diastolic VBP	0.011	0.866
Start-of-week waking systolic HBP	0.256	0.0015 *
Start-of-week waking diastolic HBP	0.087	0.448
End-of-week waking systolic HBP	0.341	<0.0001 *
End-of-week waking diastolic HBP	0.087	0.206
Start-of-week pre-HD systolic BP	0.289	<0.0001 *
Start-of-week pre-HD diastolic HDBP	0.087	0.291
End-of-week post-HD systolic HDBP	0.240	<0.0001 *
End-of-week post-HD diastolic HDBP	−0.005	0.948
WABP	0.311	<0.0001 *

LVMI, left ventricular mass index; HD, hemodialysis; hANP, human atrial natriuretic peptide; hsCRP, high-sensitivity C-reactive protein; TSat, transferrin saturation; VBP, blood pressure at the time of visit to the hospital on non-HD day; HBP, home blood pressure; HDBP, blood pressure before or after HD session; WABP, weekly averaged blood pressure. *: $p < 0.05$.

Table 5. Adjusted hazard ratio for past CV events.

Variables	HR (95% CI)	*p* Value
Model 1		0.0388
Male sex	4.39 (0.88,21.78)	0.0399
LVMI (>156 g/m^2)	2.86 (0.99,8.19)	0.0479
VBP	1.00 (0.97,1.03) (unit odds)	0.9937
Model 2		0.0718
Male sex	-	-
LVMI (>156 g/m^2)	-	-
WABP	-	-
Model 3		0.0514
Male sex	-	-
LVMI (>156 g/m^2)	-	-
Start-of-week pre-HDBP	-	-
Model 4		0.0002
Male sex	6.58 (0.75,57.44)	0.0882
LVMI (>156 g/m^2)	4.56 (1.17,17.83)	0.0291
End-of-week post-HDBP	0.94 (0.91,0.98) (unit odds)	0.0031

CV, cardiovascular; HR, hazard ratio; CI, confidence interval; LVMI, left ventricular mass index; VBP, blood pressure at the time of visit to the hospital on non-HD day; WABP, weekly averaged blood pressure; HDBP, blood pressure before or after HD session.

4. Discussion

In this prospective observational study, we found an association between increased CV events and LVMI > 156 g/m^2 with DM after performing multivariate regression analysis. End-of-week post-HD systolic BP significantly decreased 12 months after HD initiation. We also found that pre-HDBP at the start of the week, post-HDBP at the end of the week, and WABP were independently associated with LVMI on univariate regression analysis of follow-up. Multiple BP measurements taken before and after dialysis and during dialysis were reconfirmed to be the most accurate assessment format.

With regard to monitoring BP, the time point at which the BP is measured should be defined clearly. Moreover, because the BP of HD patients varies with each HD session as a result of loss of excess fluid, BP fluctuation should be considered. It is important not to limit BP evaluation to only one measurement, such as before dialysis or after dialysis; rather, multiple measurements of BP should be performed and then averaged. In this study, the number of DM cases was significantly higher in patients with CVE than in those without CVE. VBP was significantly higher in patients with LVMI > 156 g/m^2 than in those with LVMI < 156 g/m^2. The results of this study indicated that patients with low blood pressure after weekend dialysis showed an increase in past CV events. A previous study reported that systolic blood pressure < 110 mmHg and DBP < 70 mmHg were independent risk factors of CV events and all-cause mortality [20], which supports our findings. A recent study that performed sensitivity analysis using LVH and RWT separately showed that LVH but not RWT was associated with higher cardiorenal risk [21]. In this study, LVMI and RWT were not significantly different between patients with CV and those without CV events. From the pathophysiologic standpoint, an increase in afterload (i.e., arterial hypertension or an increase in large arteries stiffness) can induce concentric LVH, whereas volume overload (i.e., anemia and hypervolemic states) leads to eccentric LVH [22].

We observed that hemoglobin levels differed at 12 months after initiating HD, but hemoglobin was not an independently associated factor with LVMI. A further analysis indicated that hemoglobin levels were correlated with end-of-week post-HD systolic BP. A previous study on the independent effect of BMI reported a greater effect on LVH in women than in men [23]. There are no new discoveries about LVMI and BMI in this study.

In our study, albumin was one of the factors associated with LVMI 12 months after HD initiation. This finding is consistent with other studies showing that serum albumin is negatively correlated with LVMI [18]. hsCRP, a marker of inflammation, is also correlated with a high rate of major CV events and has been described as an independent predictor of LVMI in patients with CKD [24]. The present findings support previous observations.

Our study has some limitations. First, it was not double-blinded, and there was no control group in this study. Second, as an observational study, the findings established an association rather than a causal relationship among systolic BP, hemoglobin, and LVH. Third, the study did not evaluate patients' compliance with home ambulatory BP monitoring or compliance with antihypertensive medication regimens. Furthermore, we did not apply highly precise and reliable cardiac magnetic resonance imaging, the new gold standard for measuring LVMI [25].

Nevertheless, the findings of our study clearly defined the relative role of LVM with diabetic kidney disease in the risk assessment of patients with ESKD. Echocardiography may aid in the identification of a group at higher risk of developing CV events in HD patients. While assessing LVMI after the initiation of HD may be difficult, the treatment of hypertension, overhydration based on hANP, and anemia after dialysis initiation may reduce the progression of LVH and help identify patients at high risk for cardiovascular events before and after HD initiation.

Author Contributions: Conceptualization, H.I. (Hiroaki Io); Methodology, H.I. (Hiroaki Io) and Y.T.; Software, H.I. (Hiroaki Io); Validation, H.I. (Hiroaki Io), Y.T., and Y.S.; Formal analysis, H.I. (Hiroaki Io); Investigation, H.I. (Hiroaki Io), H.I. (Hiroyuki Inoshita), and M.I.; Resources, H.I. (Hiroaki Io), H.I. (Hiroyuki Inoshita), and M.I.; Data curation, H.I. (Hiroaki Io), J.N., H.I. (Hiroyuki Inoshita), and M.I.; Writing—original draft preparation, H.I. (Hiroaki Io); Writing—review and editing, Y.T. and Y.S.; Visualization, H.I. (Hiroaki Io), H.I. (Hiroyuki Inoshita), and M.I.; Supervision, H.I. (Hiroyuki Inoshita) and M.I.; Project administration, H.I. (Hiroaki Io), Y.T., and Y.S.; Funding acquisition, H.I. (Hiroaki Io), Y.T., and Y.S. All authors have read and agreed to the published version of the manuscript.

Funding: This research received no external funding.

Acknowledgments: We thank the nephrologists (Masako Furukawa, Kozue Okumura, Mayumi Matsumoto, Nao Nohara, Reo Kanda, and Keiichi Wakabayashi, Department of Nephrology, Juntendo University Faculty of Medicine, Tokyo, Japan) and the Juntendo University Hospital for their collaboration in this study. Editorial support in the form of medical writing based on the authors' detailed directions, collating author comments, copyediting, fact-checking, and referencing was provided by Cactus Communications. We received statistical analysis support from Amy Information Planning Co. Ltd.

Conflicts of Interest: The authors declare no conflict of interest.

References

1. Saran, R.; Li, Y.; Robinson, B.; Ayanian, J.; Balkrishnan, R.; Bragg-Gresham, J.; Chen, J.T.; Cope, E.; Gipson, D.; He, K.; et al. US Renal Data System 2014 Annual Data Report: Epidemiology of kidney disease in the United States. *Am. J. Kidney Dis.* **2015**, *66*, Svii, S1–S305. [CrossRef] [PubMed]
2. Saran, R.; Robinson, B.; Abbott, K.C.; Bragg-Gresham, J.; Chen, X.; Gipson, D.; Gu, H.; Hirth, R.A.; Hutton, D.; Jin, Y.; et al. US Renal Data System 2019 Annual Data Report: Epidemiology of kidney disease in the United States. *Am. J. Kidney Dis.* **2020**, *75*, Svi–Svii. [CrossRef] [PubMed]
3. Foley, R.N.; Curtis, B.M.; Randell, E.W.; Parfrey, P.S. Left ventricular hypertrophy in new hemodialysis patients without symptomatic cardiac disease. *Clin. J. Am. Soc. Nephrol.* **2010**, *5*, 805–813. [CrossRef] [PubMed]
4. Di Lullo, L.; Gorini, A.; Russo, D.; Santoboni, A.; Ronco, C. Left ventricular hypertrophy in chronic kidney disease patients: From pathophysiology to treatment. *Cardiorenal. Med.* **2015**, *5*, 254–266. [CrossRef]
5. Glassock, R.J.; Pecoits-Filho, R.; Barberato, S.H. Left ventricular mass in chronic kidney disease and ESRD. *Clin. J. Am. Soc. Nephrol.* **2009**, *1*, S79–S91. [CrossRef]
6. Badve, S.V.; Palmer, S.C.; Strippoli, G.; Roberts, M.A.; Teixeira-Pinto, A.; Boudville, N.; Cass, A.; Hawley, C.M.; Hiremath, S.S.; Pascoe, E.M.; et al. The validity of left ventricular mass as a surrogate end point for all-cause and cardiovascular mortality outcomes in people with CKD: A systematic review and meta-analysis. *Am. J. Kidney Dis.* **2016**, *68*, 554–563. [CrossRef]
7. Nohara, N.; Io, H.; Matsumoto, M.; Furukawa, M.; Okumura, K.; Nakata, J.; Shimizu, Y.; Horikoshi, S.; Tomino, Y. Predictive factors associated with increased progression to dialysis in early chronic kidney disease (stage 1–3) patients. *Clin. Exp. Nephrol.* **2016**, *20*, 740–747. [CrossRef]
8. Miskulin, D.C.; Gassman, J.; Schrader, R.; Gul, A.; Jhamb, M.; Ploth, D.W.; Negrea, L.; Kwong, R.Y.; Levey, A.S.; Singh, A.K.; et al. BP in dialysis: Results of a pilot study. *J. Am. Soc. Nephrol.* **2018**, *29*, 307–316. [CrossRef]
9. Io, H.; Matsumoto, M.; Okumura, K.; Sato, M.; Masuda, A.; Furukawa, M.; Nohara, N.; Tanimoto, M.; Kodama, F.; Hagiwara, S.; et al. Predictive factors associated with left ventricular hypertrophy at baseline and in the follow-up period in non-diabetic hemodialysis patients. *Semin. Dial.* **2011**, *24*, 349–354. [CrossRef]
10. Shafi, T.; Waheed, S.; Zager, P.G. Hypertension in hemodialysis patients: An opinion-based update. *Semin. Dial.* **2014**, *27*, 146–153. [CrossRef]
11. Moriya, H.; Oka, M.; Maesato, K.; Mano, T.; Ikee, R.; Ohtake, T.; Kobayashi, S. Weekly averaged blood pressure is more important than a single-point blood pressure measurement in the risk stratification of dialysis patients. *Clin. J. Am. Soc. Nephrol.* **2008**, *3*, 416–422. [CrossRef] [PubMed]
12. Merchant, A.; Wald, R.; Goldstein, M.B.; Yuen, D.; Kirpalani, A.; Dacouris, N.; Ray, J.G.; Kiaii, M.; Leipsic, J.; Kotha, V.; et al. Relationship between different blood pressure measurements and left ventricular mass by cardiac magnetic resonance imaging in end-stage renal disease. *J. Am. Soc. Hypertens.* **2015**, *9*, 275–284. [CrossRef] [PubMed]
13. Matsumoto, M.; Io, H.; Furukawa, M.; Okumura, K.; Masuda, A.; Seto, T.; Takagi, M.; Sato, M.; Nagahama, L.; Omote, K.; et al. Risk factors associated with increased left ventricular mass index in chronic kidney disease patients evaluated using echocardiography. *J. Nephrol.* **2012**, *25*, 794–801. [CrossRef] [PubMed]
14. McCullough, P.A.; Chan, C.T.; Weinhandl, E.D.; Burkart, J.M.; Bakris, G.L. Intensive hemodialysis, left ventricular hypertrophy, and cardiovascular disease. *Am. J. Kidney Dis.* **2016**, *68*, S5–S14. [CrossRef]
15. Masakane, I.; Nakai, S.; Ogata, S.; Kimata, N.; Hanafusa, N.; Hamano, T.; Wakai, K.; Wada, A.; Nitta, K. Annual dialysis data report 2014, JSDT Renal Data Registry (JRDR). *Renal Replace. Ther.* **2017**, *3*, 18. [CrossRef]
16. Nitta, K.; Iimuro, S.; Imai, E.; Matsuo, S.; Makino, H.; Akizawa, T.; Watanabe, T.; Ohashi, Y.; Hishida, A. Risk factors for increased left ventricular hypertrophy in patients with chronic kidney disease: Findings from the CKD-JAC study. *Clin. Exp. Nephrol.* **2018**, *23*, 85–98. [CrossRef]
17. Lang, R.M.; Bierig, M.; Devereux, R.B.; Flachskampf, F.A.; Foster, E.; Pellikka, P.A.; Picard, M.H.; Roman, M.J.; Seward, J.; Shanewise, J.S.; et al. Recommendations for chamber quantification: A report from the American Society of Echocardiography's Guidelines and Standards Committee and the Chamber Quantification Writing Group, developed in conjunction with the European Association of Echocardiography, a branch of the European Society of Cardiology. *J. Am. Soc. Echocardiogr.* **2005**, *18*, 1440–1463.

18. Moon, K.H.; Song, I.S.; Yang, W.S.; Shin, Y.T.; Kim, S.B.; Song, J.K.; Park, J.S. Hypoalbuminemia as a risk factor for progressive left-ventricular hypertrophy in hemodialysis patients. *Am. J. Nephrol.* **2000**, *20*, 396–401. [CrossRef]

19. Devereux, R.B.; Alonso, D.R.; Lutas, E.M.; Gottlieb, G.J.; Campo, E.; Sachs, I.; Reichek, N. Echocardiographic assessment of left ventricular hypertrophy: Comparison to necropsy findings. *Am. J. Cardiol.* **1986**, *57*, 450–458. [CrossRef]

20. Yamamoto, T.; Nakayama, M.; Miyazaki, M.; Matsushima, M.; Sato, T.; Taguma, Y.; Sato, H.; Ito, S. Relationship between low blood pressure and renal/cardiovascular outcomes in Japanese patients with chronic kidney disease under nephrologist care: The Gonryo study. *Hypertens. Res.* **2011**, *34*, 1106–1110. [CrossRef]

21. Paoletti, E.; De Nicola, L.; Gabbai, F.B.; Chiodini, P.; Ravera, M.; Pieracci, L.; Marre, S.; Cassottana, P.; Lucà, S.; Vettoretti, S.; et al. Associations of left ventricular hypertrophy and geometry with adverse outcomes in patients with CKD and hypertension. *Clin. J. Am. Soc. Nephrol.* **2016**, *11*, 271–279. [CrossRef] [PubMed]

22. Colucci, W.S.; Braunwald, E. Pathophysiology of heart failure. In *Heart Disease. A Textbook of Cardiovascular Medicine*, 6th ed.; Braunwald, E., Zipes, D.P., Libby, P., Eds.; WB Saunders Company: Philadelphia, PA, USA, 2001; pp. 503–533.

23. De Simone, G.; Devereux, R.B.; Chinali, M.; Roman, M.J.; Barac, A.; Panza, J.A.; Lee, E.T.; Howard, B.V. Sex differences in obesity-related changes in left ventricular morphology: The Strong Heart Study. *J. Hypertens.* **2011**, *29*, 1431–1438. [CrossRef]

24. Cottone, S.; Nardi, E.; Mulè, G.; Vadalà, A.; Lorito, M.C.; Riccobene, R.; Palermo, A.; Arsena, R.; Guarneri, M.; Cerasola, G. Association between biomarkers of inflammation and left ventricular hypertrophy in moderate chronic kidney disease. *Clin. Nephrol.* **2007**, *67*, 209–216. [CrossRef]

25. Agabiti-Rosei, E.; Muiesan, M.L.; Salvetti, M. New approaches to the assessment of left ventricular hypertrophy. *Ther. Adv. Cardiovasc. Dis.* **2007**, *1*, 119–128. [CrossRef] [PubMed]

Publisher's Note: MDPI stays neutral with regard to jurisdictional claims in published maps and institutional affiliations.

Journal of
Clinical Medicine

Article

Hyperphosphatemia Drives Procoagulant Microvesicle Generation in the Rat Partial Nephrectomy Model of CKD

Nima Abbasian [1,*], Alison H. Goodall [1], James O. Burton [1,2], Debbie Bursnall [3], Alan Bevington [1] and Nigel J. Brunskill [1,2]

[1] Department of Cardiovascular Sciences, University of Leicester, and Leicester NIHR Cardiovascular Biomedical Research Unit, Leicester LE3 9QP, UK; ahg5@le.ac.uk (A.H.G.); jb343@le.ac.uk (J.O.B.); 57rrd1955@gmail.com (A.B.); njb18@le.ac.uk (N.J.B.)
[2] Department of Nephrology, Leicester General Hospital, Leicester LE5 4PW, UK
[3] Division of Biomedical Services, University of Leicester, Leicester LE1 7RH, UK; db79@le.ac.uk
* Correspondence: abbasian.n.174@gmail.com; Tel.: +44(0)122-384-0020

Received: 27 September 2020; Accepted: 30 October 2020; Published: 1 November 2020

Abstract: Hyperphosphatemia has been proposed as a cardiovascular risk factor, contributing to long-term vascular calcification in hyperphosphatemic Chronic Kidney Disease (CKD) patients. However, more recent studies have also demonstrated acute effects of inorganic phosphate (Pi) on endothelial cells in vitro, especially generation of pro-coagulant endothelial microvesicles (MV). Hitherto, such direct effects of hyperphosphatemia have not been reported in vivo. Thirty-six male Sprague-Dawley rats were randomly allocated to three experimental groups: (1) CKD induced by partial nephrectomy receiving high (1.2%) dietary phosphorus; (2) CKD receiving low (0.2%) dietary phosphorus; and (3) sham-operated controls receiving 1.2% phosphorus. After 14 days the animals were sacrificed and plasma MVs counted by nanoparticle tracking analysis. MVs isolated by centrifugation were assayed for pro-coagulant activity by calibrated automated thrombography, and relative content of endothelium-derived MVs was assessed by anti-CD144 immunoblotting. When compared with sham controls, high phosphorus CKD rats were shown to be hyperphosphatemic (4.11 ± 0.23 versus 2.41 ± 0.22 mM Pi, $p < 0.0001$) with elevated total plasma MVs (2.24 ± 0.37 versus $1.31 \pm 0.24 \times 10^8$ per ml, $p < 0.01$), showing increased CD144 expression ($145 \pm 25\%$ of control value, $p < 0.0001$), and enhanced procoagulant activity (18.06 ± 1.75 versus 4.99 ± 1.77 nM peak thrombin, $p < 0.0001$). These effects were abolished in the low phosphorus CKD group. In this rat model, hyperphosphatemia (or a Pi-dependent hormonal response derived from it) is sufficient to induce a marked increase in circulating pro-coagulant MVs, demonstrating an important link between hyperphosphatemia and thrombotic risk in CKD.

Keywords: hyperphosphatemia; chronic kidney disease; cardiovascular disease; endothelial cells; procoagulant MVs

---

1. Introduction

Renal function inversely correlates with cardiovascular mortality in humans [1]. Elevation of plasma inorganic phosphate (Pi) (hyperphosphatemia) in Chronic Kidney Disease (CKD) is thought to be an important contributor to this, partly because of Pi's effect on calcium deposition, resulting in vascular calcification [2–7]. However, effects of elevated soluble Pi, apparently independent of calcium, have also been demonstrated in vitro and in vivo, for example, direct effects on parathyroid [8] and endothelial cell (EC) dysfunction [9–13]. Endothelial effects are of particular interest because CKD patients have been shown to have elevated circulating concentrations of pro-coagulant microvesicles

(MVs) derived from endothelial cells, leading to a prothrombotic state, which may contribute to acute occlusive events [14–16]. MVs are submicron diameter vesicles shed from several cell types, notably platelets and vascular endothelial cells, following apoptosis or cellular activation [9,10,17,18]. They occur in plasma of healthy subjects, but their abundance, both from platelets and ECs, has been shown to increase in CKD patients [14]. We previously showed that applying elevated extracellular Pi concentration to cultured ECs is sufficient to trigger their rapid release [9] through direct inhibition by Pi of the phosphoprotein phosphatase PP2A in ECs, culminating in cytoskeleton disruption and MV generation [10], an effect which may explain the pro-coagulant endothelial MVs previously reported in CKD patients [14]. However, a direct pro-coagulant effect of hyperphosphatemia has not been reported in vivo.

We therefore hypothesised that hyperphosphatemia in CKD in vivo is sufficient to trigger an increase in circulating pro-coagulant endothelial MVs, and that correction of hyperphosphatemia by feeding a low phosphorus diet could correct this. The study was performed using a rat partial nephrectomy model of CKD because the dietary phosphorus loads of 1.2% by weight (needed to induce stable hyperphosphatemia) and of 0.2% by weight (needed to correct hyperphosphatemia) have previously been well defined in this model [19].

2. Materials and Methods

2.1. Rat Partial Nephrectomy Model of CKD

Scheme 1 shows a schematic overview of the study design. All surgical and experimental procedures were performed subject to project licence reference PPL P444C43C0 under the Animals (Scientific Procedures) Act (United Kingdom, 1986), and were approved by the University of Leicester Animal Welfare and Ethical Review Body. Male Sprague-Dawley (SD) rats were purchased from Charles River UK at ~140–160 g. Rats (12 in each study group) were acclimatised on normal rat diet containing 0.56% phosphorus (Test Diet Limited, BCM IPS Ltd., London, UK) for 14 days before commencing surgery. A one-stage partial nephrectomy (designated "Nx" throughout this study) was performed on 24 of the resulting ~225–250 g rats under general anaesthesia as described previously [20] but a dorsal incision was used in place of abdominal access to the kidneys. Briefly, rats were anaesthetised (3.5% isoflurane in oxygen delivered at 3 litres per min), accompanied by sub-cutaneous administration of Rimadyl (Carpofen) 4 mg/kg body weight) for post-operative analgesia. Anaesthesia was maintained during surgery using 2.5% isoflurane in oxygen at 1 litre/min. The foot withdrawal reflex was tested to confirm anaesthesia. The whole of the right kidney and approximately 0.4 g of the left kidney (mainly from the cortex) were excised. The remaining 12 rats (designated "Sham" throughout this study) were anesthetised and subjected to exposure and decapsulation of both kidneys, but no excision of tissue was performed.

After a 14 day recovery period from the surgery, in which all rats were fed on the 0.56% phosphorus diet, the 24 Nx rats were randomly allocated to experimental diets: 12 receiving high a (1.2%) phosphorus diet for 14 days (referred to as "NxH" throughout this study), and 12 receiving 14 days of low (0.2%) phosphorus diet (referred to as "NxL"). Throughout this 14-day experimental period, the Sham rats were pair fed with the NxH and NxL rats, with water ad libitum. (The NxH and NxL rats received the same weight of diet but diets differed in the phosphorus content supplied). For the last 24 h rats were housed separately in metabolic cages to allow collection of urine. Thereafter, animals were sacrificed under general anaesthesia by exsanguination via cardiac puncture. Blood was collected for biochemical and MV analysis. For isolation of MVs, blood was collected into citrated tubes and MVs were separated by differential centrifugation, as previously described [9,14].

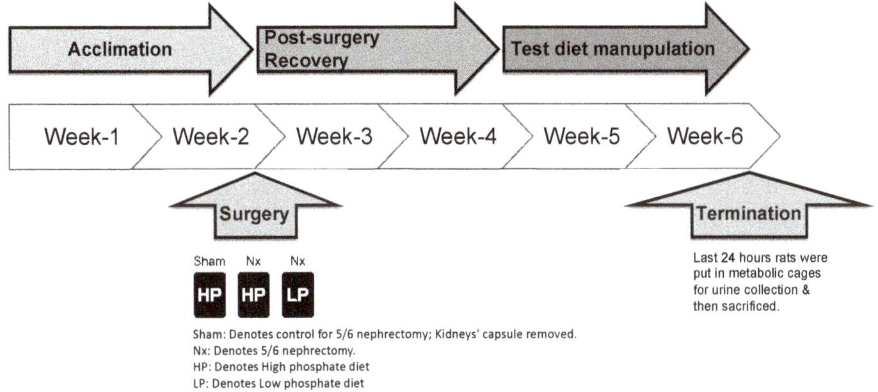

Scheme 1. Schematic overview of in vivo study design.

2.2. Blood and Urine Biochemistry

Levels of creatinine (Cr), total calcium, inorganic phosphate (Pi), and urea were measured with commercially available kits (Universal Biologicals Ltd.). Renal clearance of creatinine (expressed as mL/min/kg body weight) was calculated using the following formula:

$$((Cu/Cp) \times Vu)/BW$$

where Cu is the concentration of creatinine in urine, Cp is the concentration of creatinine in plasma, Vu is the average rate of urine production (in mL per minute) during the urine collection period, and BW is the body weight in kg.

2.3. Nanoparticle Tracking Analysis (NTA)

The number and size of the particles in the isolated MV samples was analysed by nanoparticle tracking analysis (NTA) using a NanoSight LM10 analyser with NTA software v2.2 (NanoSight Ltd., Amesbury, UK) and 90 s video capture as previously described [9,14].

2.4. Thrombin Generation Assay (TGA) Using Calibrated Automated Thrombography

The ability of isolated MVs, reconstituted in EV-free plasma pooled from 20 healthy donors, to enhance thrombin generation was determined as previously described [21] using 1×10^6 MVs per thrombin generation assay (TGA) reaction (counted by NTA as described above) by calibrated automated thrombography with Platelet-Rich Plasma Reagent (Diagnostica Stago) containing 1 pM tissue factor.

2.5. Immunoblotting

Isolated plasma MVs lysed in cell lysis buffer with 49.5 mM Tris, pH 8; 150 mM NaCl; 1% Nonidet P-40; and 1% phenylmethylsulfonyl fluoride were subjected to SDS-PAGE followed by immunoblotting. Immunoblotting was performed on nitrocellulose membranes (Amersham) followed by probing with primary mouse monoclonal antibody against CD144 (Insight Biotechnology). Polyclonal rabbit anti-mouse immunoglobulins (HRP conjugated) (DakoCytomation) were used as the secondary antibody and HRP-labelled protein was detected by chemiluminescence (ECL-Amersham). Band intensities for CD144 were quantified by ImageJ and data are presented as fold-changes compared to the sham-operated group, as the ratio of the intensity for the protein of interest (CD144)/total proteins. Cumulative intensity of total proteins detected on BioRad Mini Protean TGX stain-free gels was quantified using the stain-free gel detection facility on a BioRad ChemiDocTM Touch imaging system prior to transfer of the proteins onto the nitrocellulose membranes.

2.6. Statistical Analyses

Data are presented as bar graphs together with the mean ± SEM and were analysed using GraphPad Prism 8.0. Sample size denotes the number of rats in each study group. Differences among groups were analysed by one-way ANOVA, followed by Tukey's multiple comparisons test, for normally-distributed data, and Dunn's multiple comparisons test, for non-normally distributed data. p values < 0.05 were considered statistically significant.

3. Results

At 28 days after surgery (Scheme 1), the rat partial nephrectomy model of CKD in NxH and NxL rats showed significantly reduced renal function (assessed from creatinine clearance) (Figure 1a), and elevated serum concentrations of urea and creatinine (Figure 1b,c), when compared with that of Sham-operated control rats. Animals in all three experimental groups were successfully matched for food consumption (average cumulative food consumed over 14 days of receiving test diet; 246.12 g vs. 230.4 g vs. 257.8 g for Sham, NxH, and NxL respectively) and showed no significant difference in final body weight (Figure 1i).

As expected, renal insufficiency in NxH rats on a high phosphorus diet resulted in phosphate retention [22] shown by increased serum Pi concentration (Figure 1f) and decreased urinary Pi excretion (Figure 1g) compared with that of Sham-operated rats receiving the same dietary phosphorus intake. Feeding a low phosphorus diet to CKD rats in group NxL corrected hyperphosphatemia (Figure 1f) and abolished urinary excretion of Pi (Figure 1g). Serum calcium concentration was unaffected across the three study groups (Figure 1h).

As an increase in extracellular Pi concentration comparable with that in the NxH group (Figure 1f) had previously been shown to induce release of MVs from cultured ECs [9], the corresponding effect on MVs was investigated here in the rats' plasma (Figure 2a–c). As predicted, the MV concentration was significantly higher in the hyperphosphatemic NxH group compared with the Sham operated group, reaching statistical significance for the total MV pool (Figure 2a) and for the 10–100 nm fraction of MVs (Figure 2c). A similar upward trend was observed for the 100–1000 nm microparticle fraction (Figure 2b) but fell short of statistical significance (NxH versus Sham, $p = 0.351$, see Discussion). Correction of hyperphosphatemia (Figure 1f) by feeding the low phosphorus diet in the NxL rats abolished the increase in the MV count (Figure 2a–c). The average particle diameter showed no significant difference across the three experimental groups (Figure 2d).

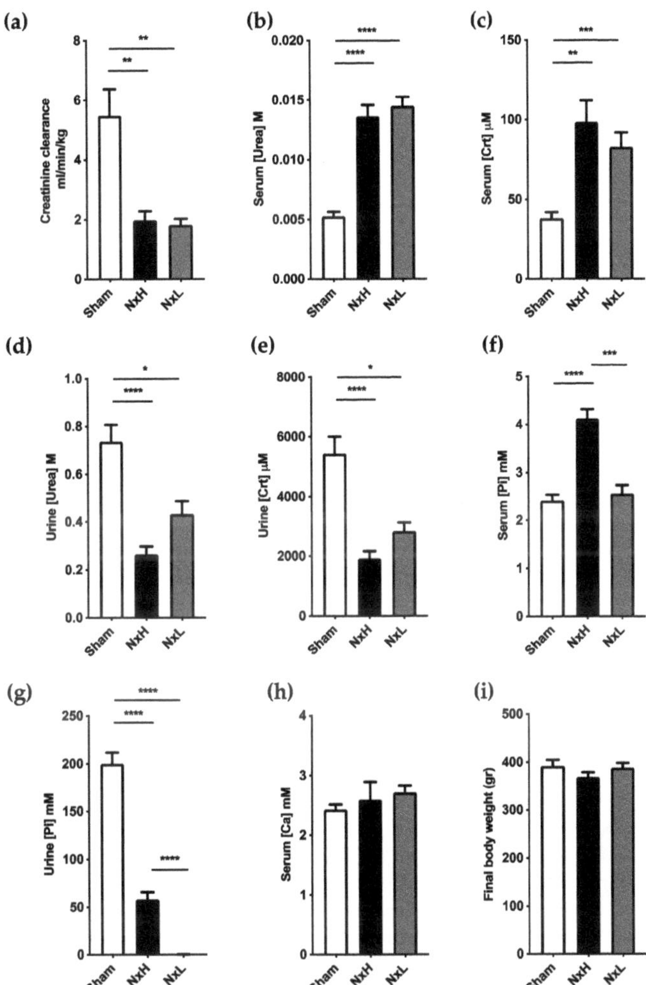

Figure 1. Development of renal failure in Chronic Kidney Disease (CKD) rats. (**a**); Creatinine clearance, (**b**); Serum (Urea), (**c**); Serum Creatinine (Crt), (**d**); Urine (Urea), (**e**); Urine (Crt), (**f**); Serum inorganic phosphate (Pi), (**g**); Urine (Pi), (**h**); Serum total Calcium (Ca), (**i**); Final body weight. Sham denotes control rats subjected to bilateral kidney decapsulation and receiving a high (1.2%) phosphorus diet for 14 days after two weeks' recovery from the surgery. NxH denotes partially nephrectomised rats receiving a high (1.2%) phosphorus diet for 14 days after two weeks' recovery from the surgery. NxL denotes partially nephrectomised rats receiving a low (0.2%) phosphorus diet for 14 days after two weeks' recovery from the surgery. $n = 12$; (12 rats in each study group), * $p < 0.05$, ** $p < 0.005$, *** $p < 0.0005$, **** $p < 0.0001$.

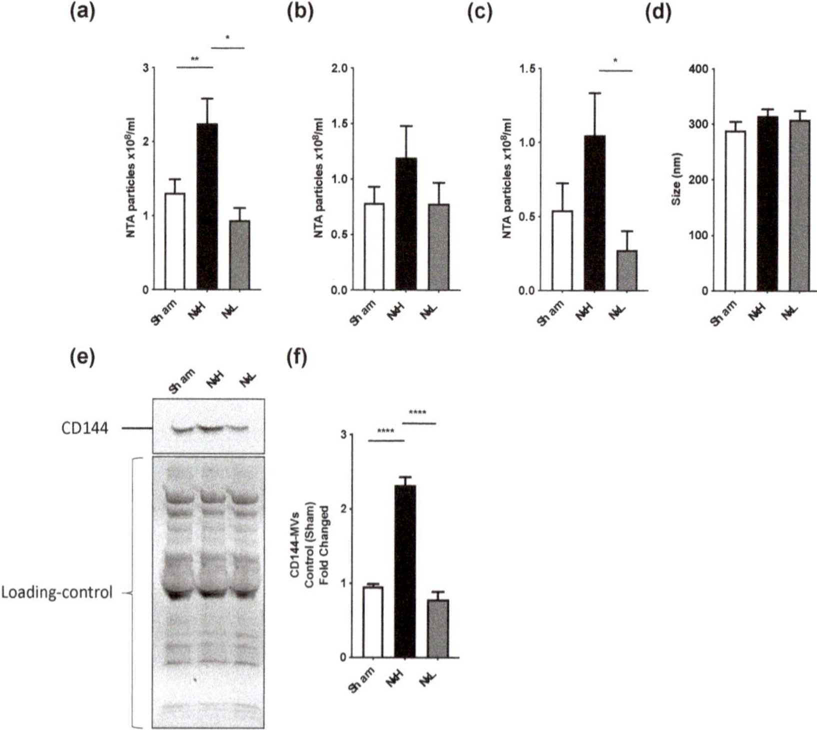

Figure 2. Hyperphosphatemia induces release of endothelial microvesicles (MVs). (**a–d**) Count and size range of MVs as measured by nanoparticle tracking analysis (NTA). (**a**); Total MVs, (**b**); MVs sized between 100–1000 nm, (**c**); 10–100 nm diameter particles, (**d**); particle size. (**e,f**) Representative Western blot results and densitometry analysis, respectively, showing the presence of CD144+ endothelium-derived MVs within isolated plasma MVs. Sham denotes control rats subjected to bilateral kidney decapsulation and receiving a high (1.2%) phosphorus diet for 14 days after two weeks' recovery from the surgery. NxH denotes partially nephrectomised rats receiving a high (1.2%) phosphorus diet for 14 days after two weeks' recovery from the surgery. NxL denotes partially nephrectomised rats receiving a low (0.2%) phosphorus diet for 14 days after two weeks' recovery from the surgery. $n = 12$; (12 rats in each study group), * $p < 0.05$, ** $p < 0.005$, **** $p < 0.0001$.

A significant contribution from endothelial vesicles to the increase in plasma MVs in the NxH rats was confirmed by immunoblotting for the endothelial marker CD144 performed on lysates from isolated MVs (Figure 2e,f). As for the MV count in Figure 2a,c, the increase in the CD144 signal observed in the NxH group was abolished in the low phosphorus NxL group (Figure 2e,f).

To confirm that the Pi-induced increase in MVs exerted a net pro-coagulant effect (as previously reported for Pi-induced MVs generated from cultured endothelial cells [9]), MVs isolated by centrifugation were assayed for pro-coagulant activity by calibrated automated thrombography (Figure 3). Both the peak thrombin concentration (Figure 3b) and endogenous thrombin potential-area under the curve (ETP-AUC) (Figure 3c) were significantly elevated in the MVs derived from the NxH rats when compared with that of Sham-operated controls. Onset of coagulation was also accelerated, shown by a reduced lag time (Figure 3d). All of these effects, which are indicative of an increased presence of procoagulant MVs, were abolished when hyperphosphatemia was corrected by feeding a low phosphorus diet in NxL rats (Figure 3b–d) and also when NxH MVs were removed by filtration (Figure 3a).

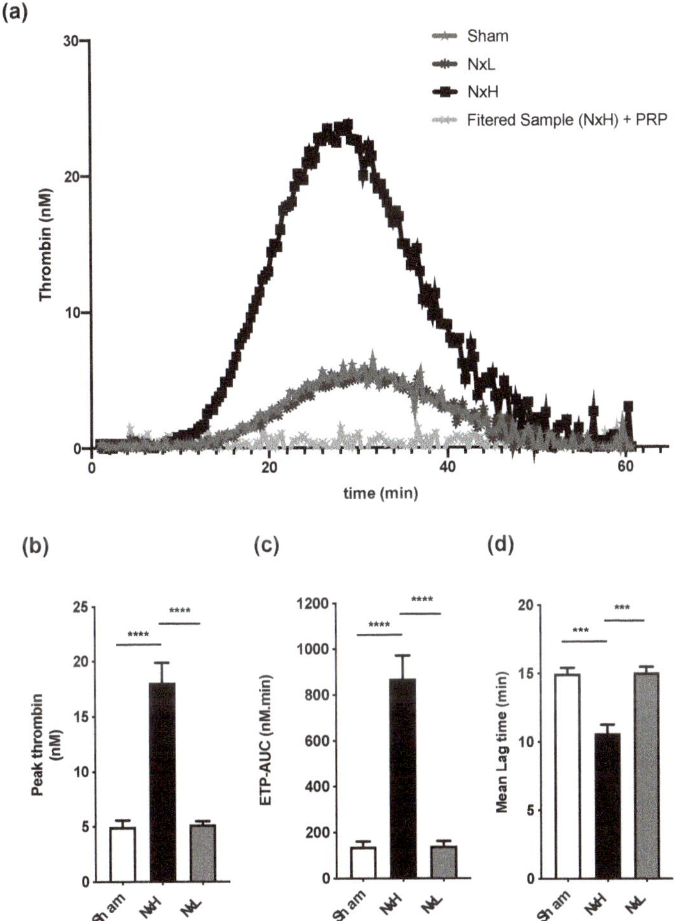

Figure 3. Calibrated automated thrombography showing the prothrombotic effects of MVs in rats' plasma. (**a**) Representative thrombin generation curves form each subject group (**b**) Peak thrombin, (**c**) Endogenous thrombin potential-area under the curve (ETP-AUC), and (**d**) Lag-time. Sham denotes control rats subjected to bilateral kidney decapsulation and receiving a high (1.2%) phosphorus diet for 14 days after two weeks' recovery from the surgery. NxH denotes partially nephrectomised rats receiving a high (1.2%) phosphorus diet for 14 days after two weeks' recovery from the surgery. NxL denotes partially nephrectomised rats receiving a low (0.2%) phosphorus diet for 14 days after two weeks' recovery from the surgery. $n = 12$; (12 rats in each study group), *** $p < 0.0005$, **** $p < 0.0001$.

4. Discussion

4.1. Contribution of Hyperphosphatemia to Pro-Coagulant MV Load in CKD

It has previously been shown that the concentration of cell-derived MVs circulating in plasma is significantly elevated in patients with CKD [14,15,23–25], that these particles are principally of endothelial, platelet, and monocyte origin, that they exert a potent pro-coagulant effect [14], and that in vitro elevated extracellular Pi concentration triggers rapid release of MVs from cultured ECs [9]. The data in the present study now provide a key advance in this field, supporting the hypothesis that in CKD it is hyperphosphatemia that triggers release of pro-coagulant MVs into circulation. It was also shown that hyperphosphatemia increased the circulating concentration of endothelial MVs (although to what extent these endothelial MVs account for the increased pro-coagulant effect remains to be shown in future work). As feeding a low phosphorus diet to CKD rats brought serum Pi concentration down to the level observed in rats with intact kidneys, and restored MV numbers and coagulation state to control level, these effects of CKD were almost entirely attributable to Pi, possibly through a direct effect of Pi ions on endothelial cells as previously described in vitro [9]. It might be argued that the effects of high versus low phosphorus diet observed in the NxH versus NxL CKD rats in Figures 1–3 arose instead from some indirect effect of dietary phosphorus (e.g., a hormonal signal from the gastro-intestinal tract to endothelium). However, feeding the same high phosphorus diet to sham control rats with intact kidneys, which yielded no hyperphosphatemia (Figure 1f), exerted no such effects on MVs, suggesting that the effects reported here are an action of elevated plasma Pi on endothelium. It should be emphasized, however, that the possibility of an indirect hormonal effect of hyperphosphatemia on endothelium (for example, a Pi-induced change in the circulating concentration of phosphatonins, such as parathyroid hormone or FGF23, which may then act on the endothelial cells) has not been ruled out, and studies of such factors would be an interesting subject for future research in this field.

4.2. Role of Endothelial Versus Platelet-Derived MVs

In this study hyperphosphatemia induced a clear increase in the total MV pool and the population in the size range 10–100 nm. Furthermore immunoblotting for endothelium-specific membrane protein CD144 in MV preparations from the NxH group of hyperphosphatemic CKD rats confirmed the increased abundance of endothelial MVs (Figure 2e,f). However, unlike our earlier in vitro study in cultured endothelial cells [9], in which 2.5 mM Pi triggered release of endothelial MVs in the microparticle size range of 100–1000 nm diameter, here, an apparent increase was seen in this range but failed to reach statistical significance (Figure 2b). It is important to emphasise, however, that a key difference between the earlier in vitro study and the present in vivo study is that in plasma in vivo the majority (typically 60%) of the particles detected in the ~100–1000 nm microparticle size range are of platelet origin [26], thus partly obscuring the hyperphosphatemia-induced rise in endothelial microparticles that was anticipated in this size range. It is unlikely that platelet microparticle numbers will increase in response to hyperphosphatemia. MV release from cultured endothelial cells in response to high extracellular Pi concentration has been shown to be triggered by the resulting increase in intracellular Pi concentration [9]. In contrast, in human platelets the steady-state intracellular Pi concentration is unaffected by hyperphosphatemia imposed in vivo [27]. Consistent with this observation, applying a 2.5 mM extracellular Pi load to human platelets for 90 min in our laboratory had no detectable effect on release of platelet MVs.

4.3. Clinical Implications

Evidence presented here (Figure 3) suggests that hyperphosphatemia contributes to increased thrombotic risk by triggering release of pro-coagulant endothelial MVs, consistent with previous evidence of pro-coagulant MVs in circulation in dialysis patients [14]. Venous thromboembolism (VTE), including deep vein thrombosis and pulmonary embolism (PE), are associated with CKD in patients [28], and decreasing eGFR has been shown to be an independent risk factor for VTE [28]. Furthermore, progression of CKD is associated with progression of atherosclerosis [29] and increased risk of blood clotting at the site of plug rupture.

The present study was not designed to investigate the time course of the effect of hyperphosphatemia on MV release in vivo. The effect was studied here after 2 weeks of dietary phosphorus intervention to ensure that a clear stable hyperphosphatemia developed in the partially nephrectomised rats [19]. However, if direct action of Pi on endothelial cells is a significant contributor to the effects reported here, onset of thrombotic risk in response to hyperphosphatemia is likely to be more rapid than 2 weeks. Work from our laboratory [9] and others [11,12,30] indicates that (at least in vitro) the response of ECs to elevated extracellular Pi concentration is rapid, and significant release of pro-coagulant MVs in vitro in response to 2.5 mM Pi occurs in as little as 90 min. This is a radically different time scale from the long-term exposure to high Pi required to induce vascular calcification. This implies that, in addition to the well-documented problems involved in managing the ubiquitous presence of phosphorus in diets [31–35], even relatively brief post-prandial "spikes" of hyperphosphatemia (which have been well documented in CKD [36–38]) may pose a significant risk.

5. Conclusions

In conclusion, this study is the first demonstration in a CKD model in vivo that hyperphosphatemia (or a Pi-dependent hormonal response derived from it) drives accumulation of pro-coagulant MVs, and hence identifies Pi as a clearly defined biochemical target for future measures to reduce thrombotic disease in CKD.

Author Contributions: Conceptualization, N.A., A.B., J.O.B., A.H.G., and N.J.B.; Data curation, N.A. and A.B.; Formal analysis, N.A. and A.B.; Funding acquisition, N.A., A.B., J.O.B., A.H.G., and N.J.B.; Investigation, N.A. and D.B.; Methodology, N.A. and A.B.; Project administration, N.A. and A.B.; Resources, N.A. and A.B.; Software, N.A. and A.B.; Supervision, A.B.; Validation, N.A. and A.B.; Visualization, N.A. and A.B.; Writing–original draft, N.A. and A.B.; Writing–review and editing, N.A., A.B., J.O.B., D.B., A.H.G., and N.J.B. All authors have read and agreed to the published version of the manuscript.

Funding: This study was supported by grant ref. RP30/July 2014 from Kidney Research UK.

Acknowledgments: The authors wish to thank the staff of the Division of Biomedical Services, University of Leicester for their valuable support in the animal work.

Conflicts of Interest: The authors declare no conflict of interest.

References

1. Henry, R.M.A.; Kostense, P.J.; Bos, G.; Dekker, J.M.; Nijpels, G.; Heine, R.J.; Bouter, L.M.; Stehouwer, C.D.A. Mild renal insufficiency is associated with increased cardiovascular mortality: The Hoorn Study. *Kidney Int.* **2002**, *62*, 1402–1407. [CrossRef] [PubMed]

2. London, G.M.; Guérin, A.P.; Marchais, S.J.; Métivier, F.; Pannier, B.; Adda, H. Arterial media calcification in end-stage renal disease: Impact on all-cause and cardiovascular mortality. *Nephrol. Dial. Transplant.* **2003**, *18*, 1731–1740. [CrossRef] [PubMed]

3. Blacher, J.; Guerin, A.P.; Pannier, B.; Marchais, S.J.; London, G.M. Arterial Calcifications, Arterial Stiffness, and Cardiovascular Risk in End-Stage Renal Disease. *Hypertension* **2001**, *38*, 938–942. [CrossRef] [PubMed]

4. Jono, S.; McKee, M.D.; Murry, C.E.; Shioi, A.; Nishizawa, Y.; Mori, K.; Morii, H.; Giachelli, C.M. Phosphate Regulation of Vascular Smooth Muscle Cell Calcification. *Circ. Res.* **2000**, *87*, E10–E17. [CrossRef]

5. Liu, L.; Liu, Y.; Zhang, Y.; Bi, X.; Nie, L.; Liu, C.; Xiong, J.; He, T.; Xu, X.; Yu, Y.; et al. High phosphate-induced downregulation of PPARgamma contributes to CKD-associated vascular calcification. *J. Mol. Cell Cardiol.* **2018**, *114*, 264–275. [CrossRef] [PubMed]

6. Reynolds, J.L.; Joannides, A.J.; Skepper, J.N.; McNair, R.; Schurgers, L.J.; Proudfoot, D.; Jahnen-Dechent, W.; Weissberg, P.L.; Shanahan, C.M. Human Vascular Smooth Muscle Cells Undergo Vesicle-Mediated Calcification in Response to Changes in Extracellular Calcium and Phosphate Concentrations: A Potential Mechanism for Accelerated Vascular Calcification in ESRD. *J. Am. Soc. Nephrol.* **2004**, *15*, 2857–2867. [CrossRef] [PubMed]

7. Giachelli, C.M.; Jono, S.; Shioi, A.; Nishizawa, Y.; Mori, K.; Morii, H. Vascular calcification and inorganic phosphate. *Am. J. Kidney Dis.* **2001**, *38*, S34–S37. [CrossRef]

8. Slatopolsky, E.; Brown, A.; Dusso, A. Pathogenesis of secondary hyperparathyroidism. *Kidney Int. Suppl.* **1999**, *73*, S14–S19. [CrossRef] [PubMed]

9. Abbasian, N.; Burton, J.O.; Herbert, K.E.; Tregunna, B.-E.; Brown, J.R.; Ghaderi-Najafabadi, M.; Brunskill, N.J.; Goodall, A.H.; Bevington, A. Hyperphosphatemia, Phosphoprotein Phosphatases, and Microparticle Release in Vascular Endothelial Cells. *J. Am. Soc. Nephrol.* **2015**, *26*, 2152–2162. [CrossRef]

10. Abbasian, N.; Bevington, A.; Burton, J.O.; Herbert, K.E.; Goodall, A.H.; Brunskill, N.J. Inorganic Phosphate (Pi) Signaling in Endothelial Cells: A Molecular Basis for Generation of Endothelial Microvesicles in Uraemic Cardiovascular Disease. *Int. J. Mol. Sci.* **2020**, *21*, 6993. [CrossRef]

11. Di Marco, G.S.; Hausberg, M.; Hillebrand, U.; Rustemeyer, P.; Wittkowski, W.; Lang, D.; Pavenstädt, H. Increased inorganic phosphate induces human endothelial cell apoptosis in vitro. *Am. J. Physiol. Physiol.* **2008**, *294*, F1381–F1387. [CrossRef] [PubMed]

12. Di Marco, G.S.; König, M.; Stock, C.; Wiesinger, A.; Hillebrand, U.; Reiermann, S.; Reuter, S.; Amler, S.; Köhler, G.; Buck, F.; et al. High phosphate directly affects endothelial function by downregulating annexin II. *Kidney Int.* **2013**, *83*, 213–222. [CrossRef] [PubMed]

13. Shuto, E.; Taketani, Y.; Tanaka, R.; Harada, N.; Isshiki, M.; Sato, M.; Nashiki, K.; Amo, K.; Yamamoto, H.; Higashi, Y.; et al. Dietary Phosphorus Acutely Impairs Endothelial Function. *J. Am. Soc. Nephrol.* **2009**, *20*, 1504–1512. [CrossRef]

14. Burton, J.O.; Hamali, H.A.; Singh, R.; Abbasian, N.; Parsons, R.; Patel, A.K.; Goodall, A.H.; Brunskill, N.J. Elevated Levels of Procoagulant Plasma Microvesicles in Dialysis Patients. *PLoS ONE* **2013**, *8*, e72663. [CrossRef]

15. Amabile, N.; Guérin, A.P.; Leroyer, A.; Mallat, Z.; Nguyen, C.; Boddaert, J.; London, G.M.; Tedgui, A.; Boulanger, C.M. Circulating Endothelial Microparticles Are Associated with Vascular Dysfunction in Patients with End-Stage Renal Failure. *J. Am. Soc. Nephrol.* **2005**, *16*, 3381–3388. [CrossRef] [PubMed]

16. Mörtberg, J.; Lundwall, K.; Mobarrez, F.; Wallén, H.; Jacobson, S.H.; Spaak, J. Increased concentrations of platelet- and endothelial-derived microparticles in patients with myocardial infarction and reduced renal function- a descriptive study. *BMC Nephrol.* **2019**, *20*, 71. [CrossRef]

17. Key, N.S.; Chantrathammachart, P.; Moody, P.W.; Chang, J.-Y. Membrane microparticles in VTE and cancer. *Thromb. Res.* **2010**, *125*, S80–S83. [CrossRef]

18. Batool, S.; Abbasian, N.; Burton, J.O.; Stover, C.M. Microparticles and their Roles in Inflammation: A Review. *Open Immunol. J.* **2013**, *6*, 1–14. [CrossRef]

19. Neves, K.-R.; Graciolli, F.-G.; Dos Reis, L.M.; Graciolli, R.-G.; Neves, C.-L.; Magalhães, A.-O.; Custódio, M.-R.; Batista, D.-G.; Jorgetti, V.; Moysés, R.M.; et al. Vascular calcification: Contribution of parathyroid hormone in renal failure. *Kidney Int.* **2007**, *71*, 1262–1270. [CrossRef]

20. Nuhu, F.; Seymour, A.-M.; Bhandari, S. Impact of Intravenous Iron on Oxidative Stress and Mitochondrial Function in Experimental Chronic Kidney Disease. *Antioxidants* **2019**, *8*, 498. [CrossRef]

21. Ambrose, A.R.; Alsahli, M.A.; Kurmani, S.A.; Goodall, A.H. Comparison of the release of microRNAs and extracellular vesicles from platelets in response to different agonists. *Platelets* **2018**, *29*, 446–454. [CrossRef]

22. Martin, K.J.; González, E.A. Prevention and Control of Phosphate Retention/Hyperphosphatemia in CKD-MBD: What Is Normal, When to Start, and How to Treat? *Clin. J. Am. Soc. Nephrol.* **2011**, *6*, 440–446. [CrossRef] [PubMed]

23. Dursun, I.; Poyrazoglu, H.M.; Gunduz, Z.; Ulger, H.; Yykylmaz, A.; Dusunsel, R.; Patyroglu, T.; Gurgoze, M. The relationship between circulating endothelial microparticles and arterial stiffness and atherosclerosis in children with chronic kidney disease. *Nephrol. Dial. Transplant.* **2009**, *24*, 2511–2518. [CrossRef] [PubMed]

24. Jalal, D.; Renner, B.; Laskowski, J.; Stites, E.; Cooper, J.; Valente, K.; You, Z.; Perrenoud, L.; Le Quintrec, M.; Muhamed, I.; et al. Endothelial Microparticles and Systemic Complement Activation in Patients With Chronic Kidney Disease. *J. Am. Heart Assoc.* **2018**, *7*, e007818. [CrossRef]

25. Erdbrügger, U.; Le, T.H. Extracellular Vesicles in Renal Diseases: More than Novel Biomarkers? *J. Am. Soc. Nephrol.* **2016**, *27*, 12–26. [CrossRef]

26. Mege, D.; Panicot-Dubois, L.; Ouaissi, M.; Robert, S.; Sielezneff, I.; Sastre, B.; Dignat-George, F.; Dubois, C. The origin and concentration of circulating microparticles differ according to cancer type and evolution: A prospective single-center study. *Int. J. Cancer* **2016**, *138*, 939–948. [CrossRef]

27. Challa, A.; Noorwali, A.; Bevington, A.; Russell, R.G.G. Cellular phosphate metabolism in patients receiving bisphosphonate therapy. *Bone* **1986**, *7*, 255–259. [CrossRef]

28. Wattanakit, K.; Cushman, M.; Stehman-Breen, C.; Heckbert, S.R.; Folsom, A.R. Chronic kidney disease increases risk for venous thromboembolism. *J. Am. Soc. Nephrol.* **2008**, *19*, 135–140. [CrossRef] [PubMed]

29. Valdivielso, J.M.; Rodríguez-Puyol, D.; Pascual, J.; Barrios, C.; Bermúdez-López, M.; Sánchez-Niño, M.D.; Pérez-Fernández, M.; Ortiz, A. Atherosclerosis in Chronic Kidney Disease: More, Less, or Just Different? *Arter. Thromb. Vasc. Biol.* **2019**, *39*, 1938–1966. [CrossRef]

30. Peng, A.; Wu, T.; Zeng, C.; Rakheja, D.; Zhu, J.; Ye, T.; Hutcheson, J.; Vaziri, N.D.; Liu, Z.; Mohan, C.; et al. Adverse Effects of Simulated Hyper- and Hypo-Phosphatemia on Endothelial Cell Function and Viability. *PLoS ONE* **2011**, *6*, e23268. [CrossRef] [PubMed]

31. Ma, G.; Li, Y.; Jin, Y.; Zhai, F.; Kok, F.J.; Yang, X. Phytate intake and molar ratios of phytate to zinc, iron and calcium in the diets of people in China. *Eur. J. Clin. Nutr.* **2007**, *61*, 368–374. [CrossRef] [PubMed]

32. Khokhar, S.; Fenwick, G.R. Phytate content of Indian foods and intakes by vegetarian Indians of Hisar Region, Haryana State. *J. Agric. Food Chem.* **1994**, *42*, 2440–2444. [CrossRef]

33. Sherman, R.A.; Mehta, O. Phosphorus and Potassium Content of Enhanced Meat and Poultry Products: Implications for Patients Who Receive Dialysis. *Clin. J. Am. Soc. Nephrol.* **2009**, *4*, 1370–1373. [CrossRef] [PubMed]

34. Sherman, R.A.; Mehta, O. Dietary Phosphorus Restriction in Dialysis Patients: Potential Impact of Processed Meat, Poultry, and Fish Products as Protein Sources. *Am. J. Kidney Dis.* **2009**, *54*, 18–23. [CrossRef] [PubMed]

35. Sullivan, C.; Sayre, S.S.; León, J.B.; Machekano, R.; Love, T.E.; Porter, D.; Marbury, M.; Sehgal, A.R. Effect of Food Additives on Hyperphosphatemia Among Patients With End-stage Renal Disease: A Randomized Controlled Trial. *JAMA* **2009**, *301*, 629–635. [CrossRef]

36. Isakova, T.; Ix, J.H.; Sprague, S.M.; Raphael, K.L.; Fried, L.; Gassman, J.J.; Raj, D.; Cheung, A.K.; Kusek, J.W.; Flessner, M.F.; et al. Rationale and Approaches to Phosphate and Fibroblast Growth Factor 23 Reduction in CKD. *J. Am. Soc. Nephrol.* **2015**, *26*, 2328–2339. [CrossRef]

37. Takeda, E.; Yamamoto, H.; Yamanaka-Okumura, H.; Taketani, Y. Dietary phosphorus in bone health and quality of life. *Nutr. Rev.* **2012**, *70*, 311–321. [CrossRef]

38. Moore, L.W.; Nolte, J.V.; Gaber, A.O.; Suki, W.N. Association of dietary phosphate and serum phosphorus concentration by levels of kidney function. *Am. J. Clin. Nutr.* **2015**, *102*, 444–453. [CrossRef]

Publisher's Note: MDPI stays neutral with regard to jurisdictional claims in published maps and institutional affiliations.

Journal of
Clinical Medicine

Article

Association between Multidimensional Prognostic Index and Hospitalization and Mortality among Older Adults with Chronic Kidney Disease on Conservative or on Replacement Therapy

Silvia Lai [1,*], Maria Ida Amabile [1], Sandro Mazzaferro [2], Giovanni Imbimbo [1],
Anna Paola Mitterhofer [1], Alessandro Galani [3], Filippo Aucella [4], Giuliano Brunori [5],
Paolo Menè [6], Alessio Molfino [1] and The Study Group on Geriatric Nephrology of the Italian
Society of Nephrology (SIN)

[1] Department of Translational and Precision Medicine, Sapienza University of Rome, 00185 Rome, Italy;
marida.amabile@gmail.com (M.I.A.); imbimbo.1638090@studenti.uniroma1.it (G.I.);
annapaola.mitter@uniroma1.it (A.P.M.); alessio.molfino@uniroma1.it (A.M.)

[2] Department of Cardiovascular, Respiratory, Nephrological, Anesthesiologic and Geriatric sciences,
Sapienza University of Rome, 00161 Rome, Italy; sandro.mazzaferro@uniroma1.it

[3] Department of Clinical and Experimental Sciences, University of Brescia, 25123 Brescia, Italy;
xelainalag@yahoo.it

[4] Nephrology and Dialysis, IRCCS "Casa Sollievo della Sofferenza", San Giovanni Rotondo, 71013 Foggia,
Italy; f.aucella@operapadrepio.it

[5] Division of Nephrology, Hospital "S. Chiara", APSS, 38122 Trento, Italy; giuliano.brunori@apss.tn.it

[6] Department of Clinical and Molecular Medicine, Sapienza University of Rome, 00189 Rome, Italy;
paolo.mene@uniroma1.it

* Correspondence: silvia.lai@uniroma1.it; Tel.: +39-393-384-094-031

Received: 16 October 2020; Accepted: 3 December 2020; Published: 7 December 2020

Abstract: The prevalence of renal disease is constantly increasing in older adults and a prognostic evaluation by a valid tool may play a key role in treatment management. We aimed to assess the association(s) between the multidimensional prognostic index (MPI) and both the hospitalization and mortality among older adults with renal disease. Patients with chronic kidney disease (CKD) (stage 3–5 KDOQI) and on dialysis were considered. Clinical parameters were registered at baseline and after 2 years. In all the patients, the MPI was calculated and divided into grade 0 (low risk), 1 (moderate risk), and 2 (severe risk). Hospitalizations and mortality were recorded during the follow-up and analyzed according to MPI grade. A total of 173 patients, with a median age of 76 years, on conservative (n = 105) and replacement therapy (32 patients on hemodialysis, 36 patients on peritoneal dialysis) were enrolled. Of them, 60 patients were in MPI grade 0, 102 in grade 1, and 11 in grade 2. The median duration of all the hospitalizations was 6 days and the number of deaths was 33. MPI significantly correlated with days of hospitalization (r = 0.801, p < 0.00001) and number of hospitalizations per year (r = 0.808, p < 0.00001), which was higher in MPI grade 2 compared to grade 1 (p < 0.001) and to grade 0 (p < 0.001). We found a significant association between MPI grades and mortality (p < 0.001). Our results indicate that MPI was associated with outcomes in patients with renal disease, suggesting that a multidimensional evaluation should be implemented in this clinical setting.

Keywords: multidimensional prognostic index; chronic kidney disease; hemodialysis; peritoneal dialysis; hospitalization; mortality

---

1. Introduction

The prevalence of older adults affected by chronic kidney disease (CKD) and requiring renal replacement therapy is high worldwide [1,2]. Currently, assessing functional, cognitive, and nutritional status among older CKD patients appears clinically relevant to stratify the risk to develop end-stage renal disease, even more than using the estimated glomerular filtration rate (eGFR) and proteinuria alone [3,4]. The prognostic evaluation of older adults with CKD is essential for physicians to identify the most appropriate clinical decision-making process for the management, treatment, and prevention of complications as well as to have a realistic expectation for patients and their family members. Studies support the idea that multidimensional assessment represents an important aspect in predicting short- and long-term all-cause mortality in older CKD patients [5–7]. This is crucial also in considering the high prevalence of frailty in this population negatively impacting on several outcomes [8].

In this light, the multidimensional prognostic index (MPI) has been found to predict mortality in patients with a variety of acute and chronic clinical conditions [4,9]. In particular, in CKD patients, the MPI was shown to be more accurate in predicting mortality when compared to the eGFR alone [7]. Data from hospital-based cohorts also indicate that using the MPI in addition to eGFR ameliorated the prediction of long-term all-cause mortality in CKD older adults [6]. Moreover, the mortality incidence rate (considering all-cause mortality) is significantly raised with the increasing of the MPI grade [6].

The MPI is based on the assessment of nutritional, cognitive, and functional status as well as on medical and social factors [5]; it is calculated from data obtained from a standardized comprehensive geriatric assessment (CGA), including six different domains such as activities of daily living (ADL), instrumental activities of daily living (IADL), short portable mental status questionnaire (SPMSQ), mini nutritional assessment (MNA), Exton-Smith score (ESS), and cumulative index rating scale (CIRS) in addition to information on medication history and cohabitation [5,6]. All these factors might negatively affect patient outcomes including hospitalization and its duration not only in CKD on conservative management but also during replacement therapy.

For this reason, we aimed to assess the association(s) of the MPI over time with the number of hospitalizations, days of hospitalization, and mortality among patients with CKD aged ≥65 years on conservative and replacement therapy, particularly hemodialysis (HD) and peritoneal dialysis (PD).

2. Materials and Methods

The study protocol was approved by the Local Clinical Research Ethics Committee (Sapienza University—Azienda Policlinico Umberto I, Rome, Italy—prot. n. 2517/15). The study conforms to the principles outlined in the Declaration of Helsinki and later amendments and we obtained a written informed consent by each patient before the enrollment.

2.1. Study Design and Participants

We performed an observational longitudinal study on clinically stable CKD patients consecutively enrolled from March 2015 to July 2017 at the University Hospital "Policlinico Umberto I" of Rome, Sapienza University of Rome, Italy. This study included CKD patients age ≥65 years on conservative therapy (eGFR ≤ 60 mL/min, stage 3–5 KDOQI), or replacement therapy (HD or PD) for at least 3 months. Statins, antihypertensive and antiplatelet therapies, and/or therapies with calcium, calcitriol and phosphate binders were continued in all patients included in the study. We recorded the clinical history and excluded patients with acute cerebrovascular and cardiovascular events within 3 months before the study, history of malignancy, or degenerative neurological or psychiatric diseases. We did not enroll patients who were not able to sign the informed consent or refused to give consent, nor did we enroll patients with missing data to calculate the MPI. We also excluded patients who were planning at the enrollment to relocate to another nephrology unit/dialysis center within the next 6 months.

The eGFR was calculated with abbreviated modification of diet in renal disease formula [10]. Clinical and laboratory variables, including hemoglobin, serum vitamin D, intact parathyroid hormone (iPTH) albumin, electrolytes, pH and base excess, were recorded at baseline and at 12 and 24 months in all CKD patients, and among HD patients, they were registered during the middle of the week, whereas among PD patients, before the first replacement of the morning with an empty peritoneum at routine visit [11]. Finally, body weight was determined to the nearest 0.1 kg using a calibrated digital scale. Body mass index (BMI) was calculated using the formula of (weight (kg)/height^2 (m^2)).

2.2. Calculation of the MPI at Baseline

We calculated the MPI as established in previous studies, consisting as a product of the CGA [5], which included data from 6 standardized scales: ADL and IADL, exploring the functional status; SPMSQ, exploring the cognitive status; MNA, investigating nutritional status; ESS for mobility and risk of pressure sore; CIRS, for multi-morbidity assessment; in addition, the number of drugs to assess polypharmacy, and co-habitation status were recorded for a total of 63 items [4,9]. The MPI was calculated in all the participants from the integrated total scores and expressed as 3 risk classes: grade 0 = low risk (MPI value between 0 and 0.33), grade 1 = moderate risk (MPI value between 0.34 and 0.66) and grade 2 = severe risk (MPI value ranging from 0.67 to 1.00) [7,9].

2.3. Hospitalization and Mortality over the 24-Month Follow-Up

All the participants were followed up with for 24 consecutive months after the enrollment and over this period we recorded the total number of days of hospitalizations per year and the number of annual admissions as well as the number of deaths per number of patients [12].

2.4. Statistical Analysis

Data management and analysis were performed using IBM® SPSS® Statistics 20.0 for Windows® software (IBM Corporation, New Orchard Road Armonk, New York, NY, USA). The normality of variables was tested using the Shapiro–Wilk method for normal distributions. All continuous variables were expressed as mean ± standard deviation, categorical variables were expressed as number (percentage). The comparison of the data of patients, for all quantitative variables considered, was performed using non-parametric Wilcoxon test and Student's *t* test. For comparing proportions was applied chi-square test. Student's t-test or the Mann–Whitney U-test were performed to determine differences between groups. The binomial test or chi-square test was used for comparison of categorical data. Pearson's correlation was used to determine, in bivariate correlation, the relationship and the strength of association between the variables, considering all the patients together and also based on the stage of the disease for hospitalization and mortality (CKD 3, CKD 4-5 or replacement therapy, including HD and PD patients). A value of $p < 0.05$ was considered statistically significant.

3. Results

3.1. Patients' Characteristics at Baseline

We initially considered 177 patients; 2 patients refused to give consent and 2 patients were excluded because they transferred to other nephrology units during the study period, making complete data unavailable. Therefore, a total of 173 patients (107 male), with a median age of 76 (70; 80) years were consecutively included; they were affected by CKD on conservative therapy (stage 3–5 KDOQI) ($n = 105$, 72 male), on HD ($n = 32$ patients, 15 male) and on PD ($n = 36$ patients, 20 male). The patients' characteristics are shown in Table 1.

3.2. MPI Classes

The MPI score was calculated at baseline and all the participants were divided into 3 risk classes: grade 0 (low risk) = 60 patients (35%), grade 1 (moderate risk) = 102 patients (59%), and grade 2 (severe risk) = 11 patients (6%).

CKD patients on conservative therapy among moderate and severe MPI risk classes (1 and 2) were 72/105 (69%), HD patients were 10/32 (31%) and patients in PD were 31/36 (86%).

Table 1. Patient characteristics. Total patients at baseline were $n = 173$ and at month 24 were $n = 140$.

	Baseline	Month 24
Male, n (%)	107 (62)	84 (60)
Age, y	76 (70; 80)	73 (69; 79)
BMI, kg/m^2	26.4 ± 4.4	23.9 ± 3.4
Hemoglobin, g/dL	11.6 (10.6; 13)	12.6 (11.6; 13.4)
Serum creatinine, mg/dL		
All	2.1(1.8; 6.0)	2.8 (1.9; 7.2)
CKD stage 3 [*]	1.8 (1.4; 2)	2.08 (1.49; 2.5)
CKD stage 4–5 [#]	1.9 (1.8; 2.2)	2.63 (2.03; 2.9)
Replacement therapy [§]	6.3 (5.6; 7)	9 (7.53; 11)
Total serum nitrogen, mg/dL		
All	97 (69; 138)	99 (87; 131)
CKD stage 3 [*]	73 (64; 95)	89 (67; 102)
CKD stage 4–5 [#]	76.5 (58; 101)	98 (76; 115)
Replacement therapy [§]	147 (114; 179)	122 (99; 148)
eGFR, mL/min/1.73 m^2		
All	27.5 (10; 39)	21 (8.2; 30.3)
CKD stage 3 [*]	39 (35; 52)	30 (25; 39)
CKD stage 4–5 [#]	25 (22; 28)	25 (19.8; 29.3)
Replacement therapy [§]	8.7 (7.3; 10.7)	7.1 (5.5; 8.8)
pH	7.33 (7.30; 7.38)	7.37 (7.33; 7.40)
Base excess	−2.50 (−6.00; 1.00)	−1.90 (−3.50; −0.80)
Sodium, mEq/L	139 (137; 142)	140 (139; 143)
Potassium, mEq/L	4.79 ± 0.63	4.63 ± 0.62
Albumin, mg/dL		
All	3.98 (3.5; 4.1)	4.60 (4.2; 5.0)
CKD stage 3 [*]	3.89 (3.45; 4.1)	4.67 (4.4; 5.0)
CKD stage 4–5 [#]	3.8 (3.5; 4)	4.7 (4.5; 5.1)
Replacement therapy [§]	4.0 (3.8; 4.4)	3.8 (3.4; 4.2)
iPTH, pg/mL	160.5 (58.5; 165)	109.5 (76.0; 221.3)
25-OH-VitD, ng/mL	21.5 (14.4; 29.3)	14.8 (7; 20)
SBP, mmHg	130 (120; 140)	130 (120; 140)
DBP, mmHg	80 (70; 86)	80 (70; 85)

Abbreviations: BMI, body mass index; CKD, chronic kidney disease; eGFR, estimated glomerular filtration rate; iPTH, intact parathyroid hormone; SBP, systolic blood pressure; DBP, diastolic blood pressure. Median (25th; 75th) is shown for non-normally distributed variables. [*] at baseline $n = 77$ and at month 24 $n = 63$; [#] at baseline $n = 28$ and at month 24 $n = 21$; [§] at baseline $n= 68$ and at month 24 $n = 56$.

3.3. 24-Month Follow-Up and Clinical Characteristics

Among the entire cohort, we registered an average 206 hospitalization per year.

At the end of the follow-up, 33 patients died ($n = 21$ of CKD, $n = 4$ of HD, $n = 8$ of PD). Therefore, a total of 140 patients with a median age of 73 years (68.75; 79) were studied at 24 months and their characteristics are shown in Table 1.

3.4. MPI and Hospitalization

We found a significant positive correlation between MPI and total number of days of hospitalization registered over the 24-month follow-up ($r = 0.801$, $p < 0.00001$) (Figure 1) as well as between MPI and the number of hospitalizations per year ($r = 0.808$, $p < 0.00001$).

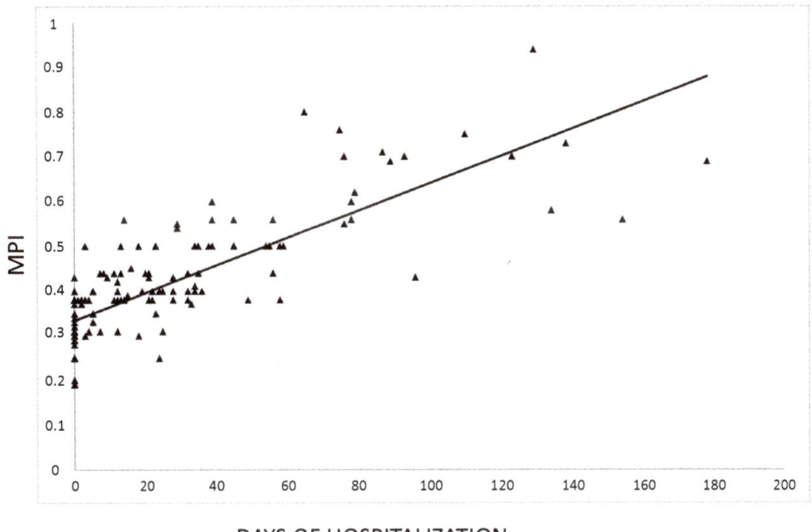

Figure 1. Correlation between the multidimensional prognostic index (MPI) score and days of hospitalization over the 24-month follow-up. ($r = 0.801$, $p < 0.00001$).

According to MPI risk classes, the mean of annual hospital admissions was different between the three MPI grades ($p < 0.001$) (Figure 2). In particular, patients in MPI risk class 2 showed a significantly higher median number of hospitalizations per year (4, IQR 3; 4) with respect to patients in MPI risk class 1 (1, IQR 1; 2) ($p < 0.0001$) and to those in MPI risk class 0 (0, IQR 0; 0) ($p < 0.0001$) as well as a higher median number of hospitalizations per year between patients in MPI risk class 1 versus those with risk 0 ($p < 0.0001$) (Figure 2). These significant differences between MPI risk classes were confirmed when considering separately CKD patients in non-replacement therapy ($p < 0.0001$), and when considering CKD patients in replacement therapy only ($p < 0.0001$).

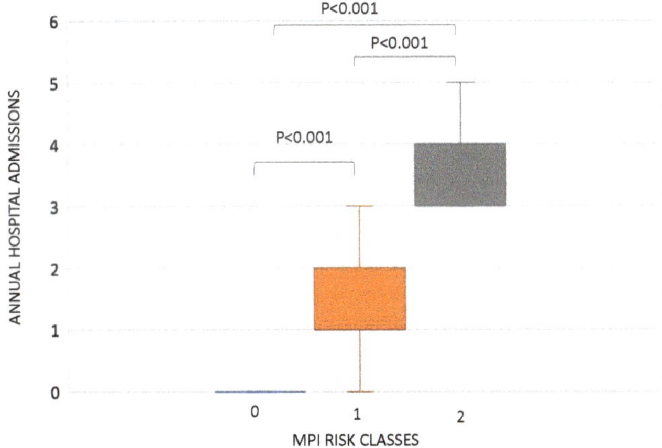

Figure 2. Differences in annual hospital admission between multidimensional prognostic index (MPI) risk class 0 (0, IQR 0; 0), MPI risk class 1 (1, IQR 1; 2) and MPI risk class 2 (4 IQR 3; 4).

Moreover, all the six domains comprised in the MPI significantly correlated with the total number of days of hospitalization and with the numbers of annual hospital admissions over the 24-month follow-up (Table 2).

Table 2. Correlations between each of the six domains of the multidimensional prognostic index (MPI) and the days of hospitalization and the number of hospitalizations per year.

	Days of Hospitalization	$n°$ of Hospitalizations per Year
Each of the 6 domains of MPI		
ADL		
r	−0.629	−0.573
p-value	<0.00001	<0.00001
IADL		
r	−0.544	−0.572
p-value	<0.00001	<0.00001
SPMSQ		
r	−0.419	−0.381
p-value	<0.00001	<0.00001
EXTON-SMITH		
r	−0.476	−0.480
p-value	<0.00001	<0.00001
CIRS		
r	0.19	0.232
p-value	0.013	0.002
MNA		
r	−0.533	−0.585
p-value	<0.00001	<0.00001

Abbreviations: ADL, activities of daily living; IADL, instrumental activities of daily living; SPMSQ, short portable mental status questionnaire; CIRS, cumulative index rating scale; MNA, mini nutritional assessment.

3.5. MPI and Mortality

We found a significant association between MPI and the number of deaths for all risk classes ($\chi^2 = 61.22$, $p < 0.0001$), and the analysis of standardized residuals (positive or negative) showed that the differences were statistically significant in each MPI risk class ($r < 1.96$ or $r > 1.96$) (Table 3).

Moreover, a significant association between MPI and the number of deaths for all risk classes was also documented when considering separately CKD patients in non-replacement therapy ($p < 0.0001$) and CKD patients in replacement therapy (HD + PD) ($p = 0.002$).

Table 3. Association between mortality (death no/yes) and multidimensional prognostic index (MPI) by risk classes. (Risk class 0 = MPI between 0 and 0.33; Risk class 1 = MPI between 0.34 and 0.66; Risk class 2 = MPI between 0.67 and 1.00).

			MPI Risk Class			Total (n)
			0	1	2	
Death	NO	Count	60	80	0	140
		Expected count	48.6	82.5	8.9	140.0
		Standardized residual	4.7	−1.0	−7.1	
	YES	Count	0	22	11	33
		Expected count	11.4	19.5	2.1	33.0
		Standardized residual	−4.7	1	7.1	
Total		Count	60	102	11	173
Person's Chi-square			Value 61.22		p-value < 0.0001	

4. Discussion

The prognostic evaluation of older adults with CKD is crucial in the decision analysis of care processes to evaluate the most appropriate management and treatment of patients with renal disease. In particular, older adults often present several comorbidities, and life expectancy is likely to be influenced by a multitude of factors. The prognosis of older adults with CKD is strongly affected by functional, cognitive, and nutritional status, psychosocial capacity, treatments, and other factors that are directly or indirectly related to the disease, suggesting the need of a prognostic tool which should be accurate in predicting mortality risk, with the objective of developing an overall comprehensive plan for treatment and follow-up [5,13]. This may be also important to decide if replacement therapy should be initiated or not. Based on this concept, the MPI was successfully used in CKD to predict all-cause mortality. Interestingly, in our study, we observed a significant association between MPI classes and outcomes not only in patients with CKD on conservative therapy (stage 3–5 KDOQI) but also in patients on HD and PD. In fact, patients on replacement therapy are known to be frail with important clinical implications [14]. In particular, among our cohort, the 86% of patients on PD were those in MPI risk classes 1 and 2.

MPI significantly correlated with the number of days of hospitalization during the 2-year follow-up, and this observation was also confirmed when analyzing the correlation between MPI and the number of hospitalizations per year. Based on these results, we believe that physicians should pay particular attention to patients within MPI risk class 2 considering that these patients were those with the highest number of hospitalizations per year. Interestingly, when considering each of the six domains comprised in the MPI, all of them significantly correlated with the total number of days of hospitalization and with hospitalizations per year. In particular, a high correlation was documented for the ADL and hospitalizations. This observation appears novel in the literature considering that the majority of the data available regarding ADL and outcomes, specifically on HD or PD, were focused on mortality [15] and not on hospitalization rate. Frailty assessment was also shown to be useful for decision-making among CKD patients on conservative care [16].

Moreover, ADL is a sensitive instrument in predicting frailty in the renal population, especially in patients in HD, and among frail patients a significantly higher rate of hospitalization was observed [17].

In this light, our data are clinically important and potentially useful for physicians in assessing the length of stay and the potential increased healthcare-related costs of patients with CKD and on replacement therapy. During our follow-up, we confirmed the significant association between MPI and mortality (number of deaths) for all the risk classes. Interestingly, all the patients identified at baseline in MPI class 2 died during the 24-month follow-up. This highlights the clinical relevance of the MPI in predicting mortality among patients with worse nutritional, cognitive, and functional status as well as with negative medical and social factors. In this view, our data are in accordance with the ones reported by others, where MPI risk classes were significantly associated with mortality in patients with renal disease [5,18]. In particular, our results may add novel information on MPI and outcomes, considering that they were obtained in a cohort of renal patients that included CKD patients at different stages and patients on dialysis, although not equally distributed in each group, followed in the same nephrology unit. This tool may be clinically useful to identify accurate and more adequate management of patients with renal disease.

Our study focused on MPI and outcomes also in patients on replacement therapy, considering the clinical relevance of predicting hospitalizations and survival in HD and PD patients. In this light, recent data showed that HD patients with frailty incurred higher healthcare costs with respect to those without frailty over a mean follow-up of 2.3 years [8]. In addition, we have previously shown that HD patients may present neurological and psychological dysfunctions that were associated with impaired quality of life [19]. Moreover, cognitive impairment was highly associated with a frail phenotype [20], resulting in increased costs and early mortality [21].

MPI calculation may also be important in PD patients to predict outcomes considering that, in this setting, we have previously found some alterations of nutritional and metabolic status, specifically among older adults (aged ≥65 years) [22].

Noteworthy is the fact that MPI includes, among others, the assessment of co-habitation status, CIRS, and polypharmacy. In particular, the use of several medications represents a clinical issue highly associated with poor prognosis during the course of several chronic diseases in the elderly [23].

In the systematic assessment of prognostic indices for all-cause mortality in older patients, the MPI has been proved to be the only selected mortality index based on a multidimensional approach, indicating the clinically relevant impact of the multidimensional derangement on the risk of mortality [5]. Interestingly, an approach that includes the assessment of several factors [5] appears challenging and may potentially add important information for clinical care in the elderly population affected by renal disease.

Our study has several limitations. We have included a highly heterogenous population consisting of patients with both moderate CKD and end-stage renal disease. The number of the participants in each group (in particular, in CKD patients in the stage 4–5 group and in the replacement group) was small, and for this reason, we could not also analyze the associations among HD and PD separately. We acknowledge that the epidemiology of the patients with CKD at earlier stage and of patients with more advanced renal disease (CKD stage 5 and dialysis) is different, likely limiting the interpretation of our results. Patients were also recruited from a single center. The implementation of MPI requires time to complete the collection of information to calculate the score and, therefore, it may limit its feasibility in clinical practice. More importantly, MPI is a tool using several items of self-reported information that may not represent an objective assessment.

5. Conclusions

In the present study, MPI was significantly associated with hospitalization and mortality over 24-month follow-up in older adults with CKD on conservative treatment or renal replacement therapy. This suggests that MPI may be clinically useful to assess prognosis in this setting and that physicians should pay attention to a multidimensional evaluation aimed at reducing patients' morbidity and mortality.

Author Contributions: Conceptualization, S.L. and A.M.; methodology, S.L., M.I.A., S.M., G.I., P.M. and A.M.; software, A.G.; formal analysis, S.L., M.I.A., G.I., A.G. and A.M.; investigation, S.L., S.M., A.P.M. and F.A.; writing—original draft preparation, S.L., M.I.A., G.I. and A.M.; writing—review and editing, S.L., S.M., G.B., P.M. and A.M. All authors have read and agreed to the published version of the manuscript.

Funding: This research received no external funding.

Conflicts of Interest: The authors declare no conflict of interest.

References

1. Pyart, R.; Evans, K.M.; Steenkamp, R.; Casula, A.; Wong, E.; Magadi, W.; Medcalf, J. The 21st UK Renal Registry Annual Report: A Summary of Analyses of Adult Data in 2017. *Nephron* **2019**, *144*, 59–66. [CrossRef]
2. United States Renal Data System. 2017 USRDS Annual Data Report: Epidemiology of kidney disease in the United States. National Institutes of Health, National Institute of Diabetes and Digestive and Kidney Diseases, Bethesda, MD, 2017. Available online: https://www.usrds.org/annual-data-report/previous-adrs (accessed on 10 October 2020).
3. Lai, S.; Muscaritoli, M.; Andreozzi, P.; Sgreccia, A.; De Leo, S.; Mazzaferro, S.; Mitterhofer, A.P.; Pasquali, M.; Protopapa, P.; Spagnoli, A.; et al. Sarcopenia and cardiovascular risk indices in patients with chronic kidney disease on conservative and replacement therapy. *Nutrition* **2019**, *62*, 108–114. [CrossRef] [PubMed]
4. Angleman, S.B.; Santoni, G.; Pilotto, A.; Fratiglioni, L.; Welmer, A.-K.; on behalf of the MPI_AGE Project Investigators. Multidimensional Prognostic Index in Association with Future Mortality and Number of Hospital Days in a Population-Based Sample of Older Adults: Results of the EU Funded MPI_AGE Project. *PLoS ONE* **2015**, *10*, e0133789. [CrossRef] [PubMed]

5. Pilotto, A.; Panza, F.; Sancarlo, D.; Paroni, G.; Maggi, S.; Ferrucci, L. Usefulness of the multidimensional prognostic index (MPI) in the management of older patients with chronic kidney disease. *J. Nephrol.* **2012**, *25*, 79–84. [CrossRef] [PubMed]

6. Pilotto, A.; Sancarlo, D.; Aucella, F.; Fontana, A.; Addante, F.; Copetti, M.; Panza, F.; Strippoli, G.F.; Ferrucci, L. Addition of the Multidimensional Prognostic Index to the Estimated Glomerular Filtration Rate Improves Prediction of Long-Term All-Cause Mortality in Older Patients with Chronic Kidney Disease. *Rejuvenation Res.* **2012**, *15*, 82–88. [CrossRef]

7. Pilotto, A.; Sancarlo, D.; Franceschi, M.; Yang, X.; D'Ambrosio, P.; Scarcelli, C.; Ferrucci, L. A multidimensional approach to the geriatric patient with chronic kidney disease. *J. Nephrol.* **2010**, *23*.

8. Sy, J.; Streja, E.; Grimes, B.; Johansen, K.L. The Marginal Cost of Frailty among Medicare Patients on Hemodialysis. *Kidney Int. Rep.* **2020**, *5*, 289–295. [CrossRef]

9. Pilotto, A.; Custodero, C.; Maggi, S.; Polidori, M.C.; Veronese, N.; Ferrucci, L. A multidimensional approach to frailty in older people. *Ageing Res. Rev.* **2020**, *60*, 101047. [CrossRef]

10. Levey, A.S.; Coresh, J.; Greene, T.; Stevens, L.A.; Zhang, Y. (Lucy); Hendriksen, S.; Kusek, J.W.; Van Lente, F.; for the Chronic Kidney Disease Epidemiology Collaboration. Using Standardized Serum Creatinine Values in the Modification of Diet in Renal Disease Study Equation for Estimating Glomerular Filtration Rate. *Ann. Intern. Med.* **2006**, *145*, 247–254. [CrossRef]

11. Lai, S.; Molfino, A.; Russo, G.E.; Testorio, M.; Galani, A.; Innico, G.; Frassetti, N.; Pistolesi, V.; Morabito, S.; Fanelli, F.R. Cardiac, Inflammatory and Metabolic Parameters: Hemodialysis versus Peritoneal Dialysis. *Cardiorenal Med.* **2015**, *5*, 20–30. [CrossRef]

12. Molfino, A.; Chiappini, M.G.; Laviano, A.; Ammann, T.; Bollea, M.R.; Alegiani, F.; Fanelli, F.R.; Molfino, A. Effect of intensive nutritional counseling and support on clinical outcomes of hemodialysis patients. *Nutrition* **2012**, *28*, 1012–1015. [CrossRef] [PubMed]

13. Volpato, S.; Bazzano, S.; Fontana, A.; Ferrucci, L.; Pilotto, A. Multidimensional Prognostic Index Predicts Mortality and Length of Stay During Hospitalization in the Older Patients: A Multicenter Prospective Study. *J. Gerontol. Ser. A Biol. Sci. Med Sci.* **2014**, *70*, 325–331. [CrossRef] [PubMed]

14. Kallenberg, M.H.; Kleinveld, H.A.; Dekker, F.W.; Van Munster, B.C.; Rabelink, T.J.; Van Buren, M.; Mooijaart, S.P. Functional and Cognitive Impairment, Frailty, and Adverse Health Outcomes in Older Patients Reaching ESRD—A Systematic Review. *Clin. J. Am. Soc. Nephrol.* **2016**, *11*, 1624–1639. [CrossRef] [PubMed]

15. McAdams-DeMarco, M.A.; Ying, H.; Olorundare, I.; King, E.A.; Haugen, C.; Buta, B.; Gross, A.L.; Kalyani, R.; Desai, N.M.; Dagher, N.N.; et al. Individual Frailty Components and Mortality in Kidney Transplant Recipients. *Transplantation* **2017**, *101*, 2126–2132. [CrossRef]

16. Villarreal, I.R.; Ortega, O.; Hinostroza, J.; Cobo, G.; Gallar, P.; Mon, C.; Herrero, J.C.; Ortiz, M.; Di Giogia, C.; Oliet, A.; et al. Geriatric Assessment for Therapeutic Decision-Making Regarding Renal Replacement in Elderly Patients with Advanced Chronic Kidney Disease. *Nephron Clin. Pract.* **2014**, *128*, 73–78. [CrossRef]

17. Lee, S.-Y.; Yang, D.H.; Hwang, E.; Kang, S.H.; Park, S.-H.; Kim, T.W.; Lee, D.H.; Park, K.; Kim, J.C. The Prevalence, Association, and Clinical Outcomes of Frailty in Maintenance Dialysis Patients. *J. Ren. Nutr.* **2017**, *27*, 106–112. [CrossRef]

18. Sancarlo, D.; Pilotto, A.; Panza, F.; Copetti, M.; Longo, M.G.; D'Ambrosio, P.; D'Onofrio, G.; Ferrucci, L.; Pilotto, A. A Multidimensional Prognostic Index (MPI) based on a comprehensive geriatric assessment predicts short- and long-term all-cause mortality in older hospitalized patients with transient ischemic attack. *J. Neurol.* **2011**, *259*, 670–678. [CrossRef]

19. Lai, S.; Molfino, A.; Mecarelli, O.; Pulitano, P.; Morabito, S.; Pistolesi, V.; Romanello, R.; Zarabla, A.; Galani, A.; Frassetti, N.; et al. Neurological and Psychological Changes in Hemodialysis Patients before and after the Treatment. *Ther. Apher. Dial.* **2018**, *22*, 530–538. [CrossRef]

20. McAdams-DeMarco, M.A.; Tan, J.; Salter, M.L.; Gross, A.L.; Meoni, L.A.; Jaar, B.G.; Kao, W.-H.L.; Parekh, R.S.; Segev, D.L.; Sozio, S.M. Frailty and Cognitive Function in Incident Hemodialysis Patients. *Clin. J. Am. Soc. Nephrol.* **2015**, *10*, 2181–2189. [CrossRef]

21. Foster, R.; Walker, S.; Brar, R.; Hiebert, B.; Komenda, P.; Rigatto, C.; Storsley, L.; Prasad, B.; Bohm, C.; Tangri, N. Cognitive Impairment in Advanced Chronic Kidney Disease: The Canadian Frailty Observation and Interventions Trial. *Am. J. Nephrol.* **2016**, *44*, 473–480. [CrossRef]

22.   Lai, S.; Amabile, M.I.; Bargagli, M.B.; Musto, T.G.; Martinez, A.; Testorio, M.; Mastroluca, D.; Lai, C.; Aceto, P.; Molfino, A.; et al. Peritoneal dialysis in older adults: Evaluation of clinical, nutritional, metabolic outcomes, and quality of life. *Medicine* **2018**, *97*, e11953. [CrossRef] [PubMed]
23.   Maher, R.L.; Hanlon, J.; Hajjar, E.R. Clinical consequences of polypharmacy in elderly. *Expert Opin. Drug Saf.* **2014**, *13*, 57–65. [CrossRef] [PubMed]

Publisher's Note: MDPI stays neutral with regard to jurisdictional claims in published maps and institutional affiliations.

Journal of
Clinical Medicine

Review

Treating Hyperuricemia: The Last Word Hasn't Been Said Yet

Elisa Russo [1], Daniela Verzola [1], Giovanna Leoncini [1,2], Francesca Cappadona [1,3], Pasquale Esposito [1,4], Roberto Pontremoli [1,2] and Francesca Viazzi [1,4,*]

[1] Department of Internal Medicine, University of Genova, Viale Benedetto XV, 6, 16132 Genova, Italy; elisa24russo@gmail.com (E.R.); daverz@libero.it (D.V.); giovanna.leoncini@unige.it (G.L.); cappadona.francesca@gmail.com (F.C.); pasqualeesposito@hotmail.com (P.E.); roberto.pontremoli@unige.it (R.P.)
[2] Internal Medicine Unit, IRCCS Ospedale Policlinico San Martino, Largo Rosanna Benzi 10, 16132 Genova, Italy
[3] Nephrologic Clinic, Sant' Andrea Hospital, Via Vittorio Veneto 197, 19121 La Spezia, Italy
[4] Nephrology Unit, IRCCS Ospedale Policlinico San Martino, Largo Rosanna Benzi 10, 16132 Genova, Italy
* Correspondence: francesca.viazzi@unige.it; Tel.: +39-0105557160

Abstract: Gout as well as asymptomatic hyperuricemia have been associated with several traditional cardiovascular risk factors and chronic kidney disease. Both in vitro studies and animal models support a role for uric acid mediating both hemodynamic and tissue toxicity leading to glomerular and tubule-interstitial damage, respectively. Nevertheless, two recent well designed and carried out trials failed to show the benefit of allopurinol treatment on kidney outcomes, casting doubts on expectations of renal protection by the use of urate lowering treatment. With the aim of providing possible explanations for the lack of effect of urate lowering treatment on chronic kidney disease progression, we will critically review results from all available randomized controlled trials comparing a urate-lowering agent with placebo or no study medication for at least 12 months and report renal clinical outcomes.

Keywords: hyperuricemia; urate lowering treatment; chronic kidney disease

Citation: Russo, E.; Verzola, D.; Leoncini, G.; Cappadona, F.; Esposito, P.; Pontremoli, R.; Viazzi, F. Treating Hyperuricemia: The Last Word Hasn't Been Said Yet. *J. Clin. Med.* **2021**, *10*, 819. https://doi.org/10.3390/jcm10040819

Academic Editor: Giacomo Garibotto
Received: 4 January 2021
Accepted: 15 February 2021
Published: 17 February 2021

Publisher's Note: MDPI stays neutral with regard to jurisdictional claims in published maps and institutional affiliations.

1. Introduction

The relationship between hyperuricemia (HU) and chronic kidney damage is bidirectional. Although a reduction in glomerular filtration rate (GFR) can precede and lead to the development of hyperuricemia, increased serum uric acid (SUA) levels per se can adversely impact renal function [1–3]. Several pathogenic mechanisms have been investigated to support the causative role of uric acid. Experimental studies show increased SUA levels might mediate kidney damage promoting innate immune response [4], inflammation [5], oxidative stress [6], activation of the renin-angiotensin aldosterone system (RAAS) [7], endothelial dysfunction [8], proliferation of vascular smooth muscle cells (VSMC) [9], resulting in glomerulosclerosis and interstitial fibrosis [10].

While population-based association studies cannot prove causation, it is fair to report that several observational studies showed elevated SUA levels are strong and independent predictors of early GFR decline and albuminuria in a very large study population with and without diabetes [11,12].

In intervention studies, xanthine oxidase inhibitors (XOIs) have been shown to reduce mean systolic and diastolic blood pressure in adolescents [13], and to improve endothelial dysfunction in specific subsets such as smokers [14] or patients with congestive heart failure [15]. While some small controlled clinical studies had previously suggested that urate lowering therapy (ULT) may retard chronic kidney disease (CKD) progression [16–18], more recent trials did not confirm a favorable effect of allopurinol on the evolution of kidney disease. In particular, in a randomized controlled trial (RCT) conducted in persons with type 1 diabetes (T1D) [19] and in the Controlled Trial of Slowing of Kidney Disease

Progression from the Inhibition of Xanthine Oxidase (CKD-Fix) [20] carried out in patients with stage 3 or 4 CKD, SUA reduction by allopurinol was unable to modify the incidence of hard renal endpoints over a long time follow up. In order to reconcile these discordant results, and to identify the characteristics of patients most likely to benefit from renal protection, in this narrative review we will critically analyze the inclusion criteria and study design of all RCTs involving the use of ULT for at least 12 months and the availability of data on renal outcome (Figure 1).

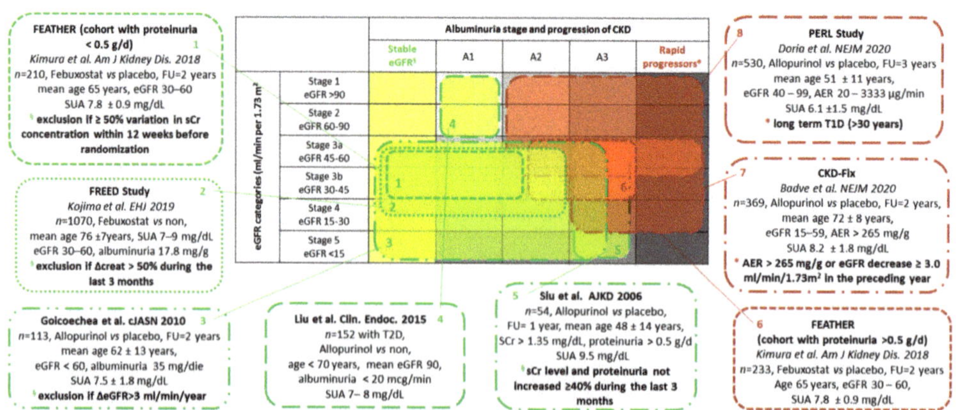

Figure 1. Schematic presentation of clinical characteristics of randomized controlled trial (RCT) patients among chronic kidney disease (CKD) categories. All available randomized controlled trials comparing a urate-lowering agent with placebo or no study medication for at least 12 months and reporting renal clinical outcomes (including kidney failure events or changes in glomerular filtration rate (GFR)) were included. The inclusion criteria (eGFR and Albuminuria categories using the NICE/KDIGO classification and the trend of kidney disease), the number of enrolled patients and baseline age, serum uric acid (SUA), eGFR and albuminuria levels were evidenced for each study. The meaning of renal disease stability or progression is specified for each study cohort. Trials with a positive or not evident renal effect attributable to urate lowering therapy (ULT) are colored with green and red, respectively. Abbreviations: CKD, chronic kidney disease; eGFR, estimated glomerular filtration rate (mL/min/1.73 m²); PERL, Preventive Early Renal Function Loss in Diabetes; CKD-Fix, controlled trial of slowing of Kidney Disease progression From the Inhibitionof Xanthine oxidase; AER, albumin excretion rate; FU, follow up; T1D, SCr, serum creatinine; type 1 diabetes; T2DM, type 2 diabetes; SUA, serum uric acid.

2. Trials with a Positive Renal Effect Attributable to ULT

With the aim of selecting the best quality studies, we decided to include in this analysis only the RCTs that investigated the renal effect of a ULT compared to placebo or no treatment for at least 12 months. Therefore, despite the rising interest in the new anti-hyperuricemia drugs, studies analyzed in this review cannot include RCTs on uricosurics drugs and are forcibly limited to those on XOis (the old Allopurinol and the newer non-purine selective XOi, Febuxostat). Topiroxostat, a new XOi approved for therapeutic use only in Japan, demonstrated a renoprotective effect by attenuating the reduction in eGFR in patients with diabetic nephropathy [21] and by inducing a 30% change of ACR in patients with renal impairment [22]. Unfortunately, RCTs on renal outcomes with Topiroxostat have a follow-up of no more than 28 weeks and therefore we did not include them in the present analysis.

We found that four studies were able to demonstrate a protective role of ULT [16–18,23,24] two studies were not [19,20] and one demonstrated a favorable effect on renal outcomes only in the subgroup of patients without proteinuria or with better initial kidney function [25]. As can be seen from Figure 1, the studies that demonstrated a protective effect of ULT were all characterized by similar inclusion criteria that guaranteed a substantial stability of renal function in the previous months [16,18,24,25] or a preserved eGFR above

60 mL/min at baseline [23]. Frequently they included patients with very low albumin excretion rate [23,24] and the presence of higher albuminuria levels judged a worse prognosis as success for the subgroup of FEATHER patients with proteinuria >0.5 g/day [25]. While some of these had a very small sample size [16,18,23] the FREED [24] is the largest RCTs available on this topic. These studies benefiting from ULT treatment include trials with both Allopurinol or Febuxostat vs. placebo or no study medication. Although Febuxostat can boast some preliminary data that suggest a greater renoprotective power than allopurinol, these findings derive from retrospective studies [26,27] or with a very short follow-up period [28].

3. PERL Study

The Preventing Early Renal Loss in Diabetes (PERL) trial enrolled 530 patients with T1D, SUA levels $\geq$ 4.5 mg/dL, mild to moderate increase in urine albumin excretion and eGFR 45–100 mL/min/1.73 m^2 or significant GFR loss (>3 mL/min/1.73 m^2/year) in the previous 3–5 years [19]. At baseline, the mean age was 51 years in the allopurinol and 52 years in the placebo group; the eGFR values were 75 and 74 mL/min/1.73 m^2, respectively. Mean SUA decreased in the allopurinol group from 6.1 at baseline to 3.9 mg/dL during treatment whereas it remained at 6.1 mg/dL in the placebo group. Despite this sustained 36% reduction in SUA, the eGFR decreased at similar rates in the two treatment groups.

Several aspects that characterize the population recruited in this study could suggest that allopurinol therapy would hardly be able to confer renal protection to these patients. First of all, the PERL study includes patients with SUA $\geq$4.5 mg/dL, which is a very low cut-off to think that modulation of SUA may have a decisive impact on the progression of kidney disease in these patients. While results were similarly neutral in pre-specified subgroup analyses based on SUA levels ($\leq$6.0 vs. >6.0 mg/dL), in the two trials by Siu and Goicoechea [16,17] showing favorable results in the arms randomized to allopurinol, the average SUA at baseline was significantly higher (9.5 and 6.7 mg/dL, respectively).

Due to the well-known relationship between low eGFR and increased SUA levels, at least in part imputable to a decreased urate excretion rate, hyperuricemia and gout are frequently observed in patients with CKD [29,30]. Therefore, in a population of patients with CKD 2 and 3 with a median urinary albumin excretion rate in the range of microalbuminuria as those recruited in the PERL study, the level of SUA that would be expected to impact the progression of renal damage is higher than that observed in this study population.

Furthermore, patients included in the study showed a very long duration of diabetes. The 530 patients experienced a 34.6 years' length of T1D, which obviously affected the natural history of their kidneys. The hyperfiltration distressing each individual nephron of these patients has inevitably triggered a vicious process leading to a progressive loss of nephronic mass that cannot be effectively countered with the ULT.

Furthermore, this population is largely represented by fast progressor patients as confirmed by the high slope of eGFR (about 2.5–3 mL/min/year) observed despite minimum or no renal damage at baseline. The FEATHER study provides a hint that febuxostat is presumably more effective for patients with less kidney damage such as those without proteinuria and those for whom serum creatinine concentration was lower than the median [25]. In a larger randomized study with a longer follow-up of about three years, Febuxostat showed a 25% reduction in the primary outcome mainly sustained by a reduced proportion of patients with a progression in albuminuria [24]. Results of the FREED study further support the view that renal protection may be more evident in the early stages of disease. In fact, in the latter study mean eGFR value was significantly better (55 mL/min, RAC 17 mg/g) than that recorded in the FEATHER study (eGFR 45 mL/min and RAC 120 mg/g). The potential of the protective effect of XOis seems to be supported by the very same data from PERL, wherein patients with normoalbuminuria seem to benefit more from treatment with allopurinol as compared to patients with a higher albumin excretion rate (although without reaching statistical significance).

In addition, the mean glycated hemoglobin level was 8.2 ± 1.3%, which means that a subset of patients was not at the target. As previously reported, severe hyperglycaemia may induce a reduction in SUA levels due to the uricosuric effect of sodium-glucose co-transporter-9 (SGLT9) activation secondary to increased glucose traffic in the tubular lumen. This mechanism leads to a J-shaped relationship between SUA and glycosylated hemoglobin, making it more complex to understand the relationship between the pharmacological modulation of urate and kidney damage.

Moreover, in the PERL study, blood pressure values were, on average, at target. This could account for the lack of nephroprotection by ULT observed in this trial. In fact, favorable cardiovascular (CV) and renal effects of SUA lowering treatment may be mediated by preventing endothelial dysfunction, vascular stiffness, and CV related events.

Finally, the large heterogeneity in the study population together with the small sample size makes it difficult to correctly interpret the data.

4. CKD Fix Study

The CKD-FIX randomized 369 patients to either allopurinol (n = 185) or placebo (n = 184) [20]. Eligible patients were adults with stage 3 or 4 CKD and albumin to creatinine ratio (ACR) ≥265 mg/g or eGFR decline rate ≥3.0 mL/min/1.73 m^2 in the preceding 12 months. Overall, baseline age (62 years), eGFR (32 mL/min/1.73 m^2), and SUA levels (8.2 mg/dL) were nearly identical in the two groups. The median urinary ACR (717 mg/g), the eGFR between 15 and 59 mL per minute per 1.73 m^2, or its decrease of at least 3.0 mL per minute per 1.73 m^2 in the preceding year, depict a figure of very high risk of progression. As discussed before, this is not the setting in which clinicians and researchers can expect the desired effect from ULT.

The trial, as clearly stated by the Authors in the limitation section, is underpowered as a result of incomplete enrollment and a high percentage of patients who discontinued the study regimen. In fact, in 20 months, recruitment reached 60% of the target number (i.e., 276 completed the trial, although the planned enrollment was 620 participants) and, based on trial logistics and funding, it was stopped by the steering committee. Moreover, during the study period, 54 patients (30%) in the allopurinol group and 45 patients (25%) in the placebo group discontinued the assigned regimen and a post hoc power calculation showed that the sample required to accommodate the discontinuation rate of 30% was 1006 patients.

An interesting hypothesis is that the pathogenic role of SUA may be different in different CKD strata. In the CKD FIX, mean SUA levels remained constant in the placebo group and decreased in the allopurinol group to 5.1 mg/dL at 12 weeks and remained at 5.3 mg/dL along the study period with a ~35% reduction substantially superimposable to that observed in the PERL study. Nevertheless, the trial did not have a serum urate level-based inclusion criterion, and this contributed to the heterogeneity of the sample study with some participants with normal and others with elevated serum urate levels at enrollment.

Once again, while the CKD-FIX Study failed to demonstrate any benefit of allopurinol on renal functional decline, this lack of effect has several possible explanations, comprising the selection of an underpowered study population with a very high risk of progression. As a matter of fact, increased SUA levels have been proven to be more predictive of kidney disease progression in the early stages of CKD and in patients without proteinuria than in patients with more severe kidney damage [31] as is the case for patients recruited in the CKD-FIX trial.

5. Findings from Reduction of Endpoints in Non-Insulin-Dependent Diabetes Mellitus with the Angiotensin II Antagonist Losartan (RENAAL) and SGLT2i Trials Support the View That SUA May Be a Modifiable Risk Factor for Renal Disease

The gold standard for renal protection in CKD are RAAS-I and SGLT2-inhibitors although this has mainly been demonstrated in subgroups of patients with clinical proteinuria and/or diabetes. Interestingly Losartan (an Angiotensin II Receptor Blocker) and gliflozins share a potentially renal protective features in that they lead to increased urinary excretion of urates. Sodium-glucose cotransporter-2 (SGLT2) inhibitors block the reabsorption of glucose at the proximal convoluted tubule and the glycosuria that results causes uric acid to be secreted into the urine. A recent meta-analysis of more than 60,000 patients showed that adults randomly assigned to receive an SGLT2 inhibitor had significantly lower SUA levels as compared to those assigned to receive a placebo or comparator medication [32]. In a longitudinal study of 300,000 adults with type 2 diabetes mellitus, a relative risk reduction in gout of nearly 40% has been observed among patients newly prescribed an SGLT2 inhibitor compared with those newly prescribed a glucagon-like peptide 1 (GLP1) agonist [33]. While the exact mechanism, in addition to glycosuria, by which SGTL2 inhibitors are thought to reduce SUA levels is not well understood, the hypothesis that this effect is part of the CV and renal protection induced by SGLT2 inhibitors is evoked by the mediation analysis undertaken in EMPAREG OUTCOME, suggesting that changes in SUA mediated ~20% to 25% of the reduction in CV death and heart failure death seen with empagliflozin [34,35].

Similar data linking a reduction of SUA levels to a positive effect on CV and renal events derives from the RENAAL Trial. Treatment with the antihypertensive drug losartan lowers SUA probably by inhibiting URAT1, leading to reduced UA reabsorption at the tubular level. In a post hoc analysis of 1342 patients with type 2 diabetes mellitus and nephropathy participating in the RENAAL, approximately one-fifth of losartan's renoprotective effect that has been proposed could be attributed to its effect on SUA [36].

6. Conclusions

While several clinical practice guidelines emphasize the usefulness of serum urate evaluation for risk stratification [37], the magnitude and importance of the SUA role in the pathogenesis of organ damage might vary and depend on the severity and duration of the underlying disease [38]. This hypothesis calls for the need of clarifying how hyperuricemia should be defined in the presence of CKD and when ULT might be prescribed for CV and renal protection in individuals with CKD.

In line with these concerns are the results of a very recent meta-analysis including 28 prospective, randomized, controlled trials assessing the effects of ULT for at least six months on CV or kidney outcomes [39]. Chen Qi et al. found that ULT was associated with the reduction of blood pressure and retardation of the decline in GFR overtime. The authors did not find benefits on clinical outcomes, including major adverse CV events, all-cause mortality, and kidney failure and once again results were conditioned by short follow-up or low quality of the trials. Also, the trials involved in this meta-analysis have significant heterogeneity related to the level of kidney function, underlying disease, and other conditions such as the usage of renin-angiotensin-aldosterone system inhibitors or significant dropout rate that could have confounded results.

In summary, although the two recently published RCTs were unable to provide the expected answers to our questions on the nephroprotective role of allopurinol, the analysis of the literature does seem to leave it open to the possibility of demonstrating the beneficial effect of ULT in future trials. Due to inclusion criteria or insufficient power, it was foreseeable that these RCTs showed no protective effect of allopurinol. Numerous data seem to suggest that the renal or vascular damage attributable to uric acid cannot regress once it has established itself. Accordingly, the presence of increased SUA levels at baseline not only predicted the development of hypertension, but also significantly blunted the decrease in blood pressure associated with lifestyle changes in children, suggesting that such children might have progressed to the irreversible phase and structural renal damage may have occurred [40]. At first, these changes are urate-dependent, but then they can trigger a self-reinforcing loop that is unresponsive to urate-lowering treatment. For this reason, patients with better preserved renal function and children might benefit more from an early ULT. Although the enrollment of patients at very low risk for progression of chronic kidney disease significantly limits the ability of a trial to show a treatment benefit, there is still much to be learned about the effect of hyperuricemia. Moreover, the health and economic burden of chronic kidney disease and its impact on increased cardiovascular risk [38] from early [41] to advanced stages of kidney function impairment [42], justify the effort to promote trials involving younger patients with earlier and less severe mild renal involvement, possibly still able to favorably respond to the reduction of uric acid levels. Therefore, while there are currently no robust data to support the routine use of pharmacotherapy for all patients with asymptomatic hyperuricemia, adequately powered, randomized, placebo-controlled trials with appropriate selection criteria are needed to determine whether specific patient groups could benefit from ULT.

Author Contributions: Conceptualization, F.V. and R.P.; validation, R.P., G.L. and P.E.; writing—original draft preparation, E.R., F.V. and R.P.; writing—review and editing, D.V., G.L. and F.C.; visualization, D.V. and P.E.; supervision, R.P.; project administration, F.V. All authors have read and agreed to the published version of the manuscript.

Funding: This research received no external funding.

Conflicts of Interest: The authors declare no conflict of interest.

References

1. Rincon-Choles, H.; Jolly, S.E.; Arrigain, S.; Konig, V.; Schold, J.D.; Nakhoul, G.; Navaneethan, S.D.; Nally, J.V., Jr.; Rothberg, M.B. Impact of Uric Acid Levels on Kidney Disease Progression. *Am. J. Nephrol.* **2017**, *46*, 315–322. [CrossRef] [PubMed]
2. Russo, E.; Drovandi, S.; Salvidio, G.; Verzola, D.; Esposito, P.; Garibotto, G.; Viazzi, F. Increased serum uric acid levels are associated to renal arteriolopathy and predict poor outcome in IgA nephropathy. *Nutr. Metab. Cardiovasc. Dis.* **2020**, *30*, 2343–2350. [CrossRef]
3. Bonino, B.; Leoncini, G.; Russo, E.; Pontremoli, R.; Viazzi, F. Uric acid in CKD: Has the jury come to the verdict? *J. Nephrol.* **2020**, *33*, 715–724. [CrossRef]
4. Braga, T.T.; Forni, M.F.; Correa-Costa, M.; Ramos, R.N.; Barbuto, J.A.; Branco, P.; Castoldi, A.; Hiyane, M.I.; Davanso, M.R.; Latz, E.; et al. Soluble Uric Acid Activates the NLRP3 Inflammasome. *Sci. Rep.* **2017**, *7*, 39884. [CrossRef]
5. Braga, T.T.; Foresto-Neto, O.; Camara, N.O.S. The role of uric acid in inflammasome-mediated kidney injury. *Curr. Opin. Nephrol. Hypertens.* **2020**, *29*, 423–431. [CrossRef] [PubMed]
6. Yang, L.; Chang, B.; Guo, Y.; Wu, X.; Liu, L. The role of oxidative stress-mediated apoptosis in the pathogenesis of uric acid nephropathy. *Ren. Fail.* **2019**, *41*, 616–622. [CrossRef] [PubMed]
7. Perlstein, T.S.; Gumieniak, O.; Hopkins, P.N.; Murphey, L.J.; Brown, N.J.; Williams, G.H.; Hollenberg, N.K.; Fisher, N.D. Uric acid and the state of the intrarenal renin-angiotensin system in humans. *Kidney Int.* **2004**, *66*, 1465–1470. [CrossRef] [PubMed]
8. Zhen, H.; Gui, F. The role of hyperuricemia on vascular endothelium dysfunction. *Biomed. Rep.* **2017**, *7*, 325–330. [CrossRef]
9. Corry, D.B.; Eslami, P.; Yamamoto, K.; Nyby, M.D.; Makino, H.; Tuck, M.L. Uric acid stimulates vascular smooth muscle cell proliferation and oxidative stress via the vascular renin–angiotensin system. *J. Hypertens.* **2008**, *26*, 269–275. [CrossRef]
10. Jing, P.; Shi, M.; Ma, L.; Fu, P. Mechanistic Insights of Soluble Uric Acid-related Kidney Disease. *Curr. Med. Chem.* **2020**, *27*, 5056–5066. [CrossRef] [PubMed]
11. De Cosmo, S.; Viazzi, F.; Pacilli, A.; Giorda, C.; Ceriello, A.; Gentile, S.; Russo, G.; Rossi, M.C.; Nicolucci, A.; Guida, P.; et al. Serum Uric Acid and Risk of CKD in Type 2 Diabetes. *Clin. J. Am. Soc. Nephrol.* **2015**, *10*, 1921–1929. [CrossRef]
12. Srivastava, A.; Kaze, A.D.; McMullan, C.J.; Isakova, T.; Waikar, S.S. Uric Acid and the Risks of Kidney Failure and Death in Individuals With CKD. *Am. J. Kidney Dis.* **2018**, *71*, 362–370. [CrossRef]
13. Feig, D.I.; Soletsky, B.; Johnson, R.J. Effect of Allopurinol on Blood Pressure of Adolescents with Newly Diagnosed Essential Hypertension. *JAMA* **2008**, *300*, 924–932. [CrossRef]
14. Guthikonda, S.; Sinkey, C.; Barenz, T.; Haynes, W.G. Xanthine Oxidase Inhibition Reverses Endothelial Dysfunction in Heavy Smokers. *Circulation* **2003**, *107*, 416–421. [CrossRef]
15. Doehner, W.; Schoene, N.; Rauchhaus, M.; Leyva-Leon, F.; Pavitt, D.V.; Reaveley, D.A.; Schuler, G.; Coats, A.J.; Anker, S.D.; Hambrecht, R. Effects of xanthine oxidase inhibition with allopurinol on endothelial function and peripheral blood flow in hyperuricemic patients with chronic heart failure: Results from 2 placebo-controlled studies. *Circulation* **2002**, *105*, 2619–2624. [CrossRef] [PubMed]
16. Siu, Y.-P.; Leung, K.-T.; Tong, M.K.-H.; Kwan, T.-H. Use of Allopurinol in Slowing the Progression of Renal Disease Through Its Ability to Lower Serum Uric Acid Level. *Am. J. Kidney Dis.* **2006**, *47*, 51–59. [CrossRef] [PubMed]

17. Goicoechea, M.; De Vinuesa, S.G.; Verdalles, U.; Ruiz-Caro, C.; Ampuero, J.; Rincón, A.; Arroyo, D.; Luño, J. Effect of Allopurinol in Chronic Kidney Disease Progression and Cardiovascular Risk. *Clin. J. Am. Soc. Nephrol.* **2010**, *5*, 1388–1393. [CrossRef]

18. Goicoechea, M.; De Vinuesa, S.G.; Verdalles, U.; Verde, E.; Macias, N.; Santos, A.; De Jose, A.P.; Cedeño, S.; Linares, T.; Luño, J. Allopurinol and Progression of CKD and Cardiovascular Events: Long-term Follow-up of a Randomized Clinical Trial. *Am. J. Kidney Dis.* **2015**, *65*, 543–549. [CrossRef] [PubMed]

19. Doria, A.; Galecki, A.T.; Spino, C.; Pop-Busui, R.; Cherney, D.Z.; Lingvay, I.; Parsa, A.; Rossing, P.; Sigal, R.J.; Afkarian, M.; et al. Serum Urate Lowering with Allopurinol and Kidney Function in Type 1 Diabetes. *N. Engl. J. Med.* **2020**, *382*, 2493–2503. [CrossRef]

20. Badve, S.V.; Pascoe, E.M.; Tiku, A.; Boudville, N.; Brown, F.G.; Cass, A.; Clarke, P.; Dalbeth, N.; Day, R.O.; De Zoysa, J.R.; et al. Effects of Allopurinol on the Progression of Chronic Kidney Disease. *N. Engl. J. Med.* **2020**, *382*, 2504–2513. [CrossRef] [PubMed]

21. Wada, T.; Hosoya, T.; Honda, D.; Sakamoto, R.; Narita, K.; Sasaki, T.; Okui, D.; Kimura, K. Uric acid-lowering and renoprotective effects of topiroxostat, a selective xanthine oxidoreductase inhibitor, in patients with diabetic nephropathy and hyperuricemia: A randomized, double-blind, placebo-controlled, parallel-group study (UPWARD study). *Clin. Exp. Nephrol.* **2018**, *22*, 860–870. [CrossRef]

22. Hosoya, T.; Ohno, I.; Nomura, S.; Hisatome, I.; Uchida, S.; Fujimori, S.; Yamamoto, T.; Hara, S. Effects of topiroxostat on the serum urate levels and urinary albumin excretion in hyperuricemic stage 3 chronic kidney disease patients with or without gout. *Clin. Exp. Nephrol.* **2014**, *18*, 876–884. [CrossRef]

23. Liu, P.; Chen, Y.; Wang, B.; Zhang, F.; Wang, D.; Wang, Y. Allopurinol treatment improves renal function in patients with type 2 diabetes and asymptomatic hyperuricemia: 3-year randomized parallel-controlled study. *Clin. Endocrinol.* **2014**, *83*, 475–482. [CrossRef] [PubMed]

24. Kojima, S.; Matsui, K.; Hiramitsu, S.; Hisatome, I.; Waki, M.; Uchiyama, K.; Yokota, N.; Tokutake, E.; Wakasa, Y.; Jinnouchi, H.; et al. Febuxostat for Cerebral and CaRdiorenovascular Events PrEvEntion StuDy. *Eur. Hear. J.* **2019**, *40*, 1778–1786. [CrossRef] [PubMed]

25. Kimura, K.; Hosoya, T.; Uchida, S.; Inaba, M.; Makino, H.; Maruyama, S.; Ito, S.; Yamamoto, T.; Tomino, Y.; Ohno, I.; et al. Febuxostat Therapy for Patients With Stage 3 CKD and Asymptomatic Hyperuricemia: A Randomized Trial. *Am. J. Kidney Dis.* **2018**, *72*, 798–810. [CrossRef]

26. Yang, A.Y. Comparison of long-term efficacy and renal safety of febuxostat and allopurinol in patients with chronic kidney diseases. *Int. J. Clin. Pharmacol. Ther.* **2020**, *58*, 21–28. [CrossRef] [PubMed]

27. Hu, A.M.; Brown, J.N. Comparative effect of allopurinol and febuxostat on long-term renal outcomes in patients with hyperuricemia and chronic kidney disease: A systematic review. *Clin. Rheumatol.* **2020**, *39*, 3287–3294. [CrossRef]

28. Sezai, A.; Soma, M.; Nakata, K.-I.; Hata, M.; Yoshitake, I.; Wakui, S.; Hata, H.; Shiono, M. Comparison of Febuxostat and Allopurinol for Hyperuricemia in Cardiac Surgery Patients (NU-FLASH Trial). *Circ. J.* **2013**, *77*, 2043–2049. [CrossRef] [PubMed]

29. Tan, V.S.; Garg, A.X.; McArthur, E.; Lam, N.N.; Sood, M.M.; Naylor, K.L. The 3-Year Incidence of Gout in Elderly Patients with CKD. *Clin. J. Am. Soc. Nephrol.* **2017**, *12*, 577–584. [CrossRef]

30. Viazzi, F.; Garneri, D.; Leoncini, G.; Gonnella, A.; Muiesan, M.; Ambrosioni, E.; Costa, F.; Leonetti, G.; Pessina, A.; Trimarco, B.; et al. Serum uric acid and its relationship with metabolic syndrome and cardiovascular risk profile in patients with hypertension: Insights from the I-DEMAND study. *Nutr. Metab. Cardiovasc. Dis.* **2014**, *24*, 921–927. [CrossRef]

31. Tsai, C.-W.; Lin, S.-Y.; Kuo, C.-C.; Huang, C.-C. Serum Uric Acid and Progression of Kidney Disease: A Longitudinal Analysis and Mini-Review. *PLoS ONE* **2017**, *12*, e0170393. [CrossRef]

32. Zhao, Y.; Xu, L.; Tian, D.; Xia, P.; Zheng, H.; Wang, L.; Chen, L. Effects of sodium-glucose co-transporter 2 (SGLT2) inhibitors on serum uric acid level: A meta-analysis of randomized controlled trials. *Diabetes Obes. Metab.* **2018**, *20*, 458–462. [CrossRef] [PubMed]

33. Fralick, M.; Chen, S.K.; Patorno, E.; Kim, S.C. Assessing the Risk for Gout with Sodium-Glucose Cotransporter-2 Inhibitors in Patients with Type 2 Diabetes: A Population-Based Cohort Study. *Ann. Intern. Med.* **2020**, *172*, 186–194. [CrossRef]

34. Fitchett, D.; Inzucchi, S.E.; Zinman, B.; Wanner, C.; Zannad, F.; Schumacher, M.; Schmoor, C.; Ohneberg, K.; Salsali, A.; Jyothis, T.G.; et al. Mediators of the Improvement in Heart Failure Outcomes With Empagliflozin in the EMPAREG OUTCOME Trial. *Circulation* **2017**, *136*, A15893.

35. Inzucchi, S.E.; Zinman, B.; Fitchett, D.; Wanner, C.; Ferrannini, E.; Schumacher, M.; Schmoor, C.; Ohneberg, K.; Johansen, O.E.; George, J.T.; et al. How Does Empagliflozin Reduce Cardiovascular Mortality? Insights from a Mediation Analysis of the EMPA-REG OUTCOME Trial. *Diabetes Care* **2017**, *41*, 356–363. [CrossRef]

36. Miao, Y.; Ottenbros, S.A.; Laverman, G.D.; Brenner, B.M.; Cooper, M.E.; Parving, H.-H.; Grobbee, D.E.; Shahinfar, S.; De Zeeuw, D.; Heerspink, H.J.L. Effect of a Reduction in Uric Acid on Renal Outcomes During Losartan Treatment: A Post Hoc Analysis of the Reduction of Endpoints in Non-Insulin-Dependent Diabetes Mellitus With the Angiotensin II Antagonist Losartan Trial. *Hypertension* **2011**, *58*, 2–7. [CrossRef] [PubMed]

37. Williams, B.; Mancia, G.; Spiering, W.; Rosei, E.A.; Azizi, M.; Burnier, M.; Clement, D.L.; Coca, A.; de Simone, G.; Dominiczak, A.F.; et al. 2018 ESC/ESH Guidelines for the management of arterial hypertension. *Eur. Heart J.* **2018**, *39*, 3021–3104. [CrossRef] [PubMed]

38. Virdis, A.; Masi, S.; Casiglia, E.; Tikhonoff, V.; Cicero, A.F.; Ungar, A.; Rivasi, G.; Salvetti, M.; Barbagallo, C.M.; Bombelli, M.; et al. Identification of the Uric Acid Thresholds Predicting an Increased Total and Cardiovascular Mortality Over 20 Years. *Hypertension* **2020**, *75*, 302–308. [CrossRef]

39. Chen, Q.; Wang, Z.; Zhou, J.; Chen, Z.; Li, Y.; Li, S.; Zhao, H.; Badve, S.V.; Lv, J. Effect of Urate-Lowering Therapy on Cardiovascular and Kidney Outcomes. *Clin. J. Am. Soc. Nephrol.* **2020**, *15*, 1576–1586. [CrossRef]
40. Viazzi, F.; Rebora, P.; Giussani, M.; Orlando, A.; Stella, A.; Antolini, L.; Valsecchi, M.G.; Pontremoli, R.; Genovesi, S. Increased Serum Uric Acid Levels Blunt the Antihypertensive Efficacy of Lifestyle Modifications in Children at Cardiovascular Risk. *Hypertension* **2016**, *67*, 934–940. [CrossRef] [PubMed]
41. Bezante, G.P.; Viazzi, F.; Leoncini, G.; Ratto, E.; Conti, N.; Balbi, M.; Agosti, S.; Deferrari, G.; Pontremoli, R.; Deferrari, L. Coronary Flow Reserve Is Impaired in Hypertensive Patients With Subclinical Renal Damage. *Am. J. Hypertens.* **2009**, *22*, 191–196. [CrossRef] [PubMed]
42. Rahman, M.; Pressel, S.; Davis, B.R.; Nwachuku, C.; Wright, J.T.; Whelton, P.K.; Barzilay, J.; Batuman, V.; Eckfeldt, J.H.; Farber, M.A.; et al. Cardiovascular Outcomes in High-Risk Hypertensive Patients Stratified by Baseline Glomerular Filtration Rate. *Ann. Intern. Med.* **2006**, *144*, 172–180. [CrossRef] [PubMed]

Journal of
Clinical Medicine

Article

Association between *MANBA* Gene Variants and Chronic Kidney Disease in a Korean Population

Hye-Rim Kim [1,†], Hyun-Seok Jin [2,†] and Yong-Bin Eom [1,3,*]

[1]   Department of Medical Sciences, Graduate School, Soonchunhyang University,
     Asan 31538, Chungnam, Korea; goa6471@naver.com
[2]   Department of Biomedical Laboratory Science, College of Life and Health Sciences, Hoseo University,
     Asan 31499, Chungnam, Korea; jinhs@hoseo.edu
[3]   Department of Biomedical Laboratory Science, College of Medical Sciences, Soonchunhyang University,
     Asan 31538, Chungnam, Korea
*   Correspondence: omnibin@sch.ac.kr; Tel.: +82-41-530-3039
†   These authors contributed equally to this work.

Abstract: Chronic kidney disease (CKD), a damaged condition of the kidneys, is a global public health problem that can be caused by diabetes, hypertension, and other disorders. Recently, the *MANBA* gene was identified in CKD by integrating CKD-related variants and kidney expression quantitative trait loci (eQTL) data. This study evaluated the effects of *MANBA* gene variants on CKD and kidney function-related traits using a Korean cohort. We also analyzed the association of *MANBA* gene variants with kidney-related traits such as the estimated glomerular filtration rate (eGFR), and blood urea nitrogen (BUN), creatinine, and uric acid levels using linear regression analysis. As a result, 14 single nucleotide polymorphisms (SNPs) were replicated in CKD ($p < 0.05$), consistent with previous studies. Among them, rs4496586, which was the most significant for CKD and kidney function-related traits, was associated with a decreased CKD risk in participants with the homozygous minor allele (CC), increased eGFR, and decreased creatinine and uric acid concentrations. Furthermore, the association analysis between the rs4496586 genotype and *MANBA* gene expression in human tubules and glomeruli showed high *MANBA* gene expression in the minor allele carriers. In conclusion, this study demonstrated that *MANBA* gene variants were associated with CKD and kidney function-related traits in a Korean cohort.

Keywords: *MANBA*; variants; chronic kidney disease; estimated glomerular filtration rate (eGFR); expression quantitative trait loci (eQTL)

Citation: Kim, H.-R.; Jin, H.-S.; Eom, Y.-B. Association between *MANBA* Gene Variants and Chronic Kidney Disease in a Korean Population. *J. Clin. Med.* **2021**, *10*, 2255. https://doi.org/10.3390/jcm10112255

Academic Editor: Giacomo Garibotto

Received: 13 April 2021
Accepted: 21 May 2021
Published: 23 May 2021

Publisher's Note: MDPI stays neutral with regard to jurisdictional claims in published maps and institutional affiliations.

1. Introduction

Chronic kidney disease (CKD) is an important health problem worldwide and increases mortality and morbidity by raising the risk of several diseases [1,2]. According to the Global Burden of Disease (GBD) study, the global prevalence of CKD in 2017 was estimated at 9.7%, with approximately 697.5 million people affected [3]. The prevalence of CKD in Korea in 2017 was 3% according to the chronic disease health statistics of the Korea Disease Control and Prevention Agency (KDCA) (https://health.cdc.go.kr/ (accessed on 13 April 2021)). In addition, the GBD study reported that fatality from CKD in 2017 was 1.2 million, the 12th primary cause of death worldwide. The common causes of impaired kidney function are diabetes, hypertension, and glomerulonephritis, which can be evaluated by the estimated glomerular filtration rate (eGFR) [4]. Previous studies have estimated the heritability of CKD to be 20–80% by measuring the contribution of genetic effects in a population of patients with CKD [5–7]. Therefore, since CKD is also affected by genetic factors, it is important to identify the risk factors associated with CKD development and genes in individuals with CKD.

A meta-analysis of genome-wide association studies (GWAS) on CKD and kidney function-related traits was performed in European, Asian, and African populations [8–10].

141

Previous studies have determined non-coding genetic variants related to CKD through GWAS but the elucidation of the underlying genes and mechanisms was limited. Therefore, a recent study performed expression quantitative trait loci (eQTL) analysis of renal glomeruli and tubular tissue and found 27 candidate genes that could be potential causes of kidney disease development [11]. Among them, *MANBA*, *PGAP3*, and *CASP9* were confirmed to have functional roles in kidney disease development in previous animal studies [12–14]. Another study reported that variants of the *MANBA* gene identified using eQTL analysis and CKD-related variants identified using GWAS analysis showed statistically significant co-localization [12]. In addition, Gu et al. demonstrated the in vivo mechanism by which the *MANBA* gene affects kidney disease [15].

The *β*-mannosidase (*MANBA*) gene, encoding lysosome *β*-mannosidase, is located on human chromosome 4q24 [16,17]. Although many genetic variants associated with CKD have been identified, correlation analysis focusing on *MANBA* gene variants that directly affect kidney diseases are rare. Therefore, in this study, the association analysis of *MANBA* gene variants with CKD and kidney function-related traits was performed in the Korean Genome and Epidemiology Study (KoGES) cohort. We found 20 single nucleotide polymorphisms (SNPs) that showed a statistically significant association with CKD and kidney function-related traits among 229 SNPs of the *MANBA* gene. In addition, rs4496586, which had the highest significance for CKD, was associated with *MANBA* gene expression in renal tubules and glomeruli. These results replicate previous studies that functionally demonstrated the association between kidney disease and the *MANBA* gene.

2. Materials and Methods

2.1. Participants

The epidemiological data used in this study were obtained from the Health Examinee (HEXA) cohort of the Korean Genome and Epidemiology Study (KoGES). From 2004 to 2013, a total of 173,208 participants aged over 40 years were recruited in the HEXA cohort. Among them, genotype data on 58,700 participants were available. A more detailed description of the HEXA cohort has been described previously [18]. The Institutional Review Board (IRB) of the Korea Disease Control and Prevention Agency (KDCA, KBN-2021-003 (26 January 2021)) and Soonchunhyang University (202012-BR-086-01 (15 December 2020)) approved the study protocol. Written informed consent was obtained from all subjects. All methods were performed in accordance with the relevant guidelines and regulations.

2.2. Basic Characteristics

The parameters measured in this study included physical measurements such as height, and weight, and biochemical measurements such as uric acid, blood urea nitrogen (BUN), and serum creatinine levels. The characteristics of the participants in this study are shown in Table 1. Body mass index (BMI) was calculated by dividing body weight (kg) by height squared (m^2). Blood samples were collected after an 8-h fast and all plasma samples were measured biochemically. Serum creatinine concentrations were measured by the Jaffe method using an automatic analyzer (Hitachi, Tokyo, Japan).

2.3. Definition of Chronic Kidney Disease

For the case-control analysis of CKD, control groups (30,813 participants) and cases (1130 participants) were classified according to the recommendations of the Kidney Disease Improving Global Outcome (KDIGO) guidelines. CKD was defined as an eGFR of <60 mL/min/1.72 m^2 and a history of renal disease. Non-CKD was defined as an eGFR of ≥90 mL/min/1.72 m^2. Known as the best estimate of kidney function, the eGFR was calculated through the creatinine-based modification of diet in renal disease (MDRD) equation. Kidney function was additionally assessed by BUN, uric acid, and serum creatinine levels.

Table 1. Characteristics of participants in the Korean population.

Characteristics	Quantitative Trait Analysis	Case-Control Analysis for CKD		
		Controls	Cases	*p*-Value *
Number of participants	58,700	30,813	1130	
Gender [men (%)]	20,293 (34.57)	9170 (29.76)	528 (46.73)	<0.001
Age (M years ± SD)	53.8 ± 8.02	52.27 ± 7.53	60.92 ± 6.91	<0.001
Height (M cm ± SD)	160.72 ± 7.93	160.10 ± 7.76	161.10 ± 8.02	<0.001
Weight (M kg ± SD)	61.89 ± 9.89	60.95 ± 9.65	64.46 ± 10.09	<0.001
BMI (M kg/m^2 ± SD)	23.67 ± 2.52	23.57 ± 2.66	24.76 ± 2.87	0.557
eGFR (mL/min/1.73 m^2)	91.22 ± 16.54	103.35 ± 11.82	53.80 ± 9.69	<0.001
BUN (mg/dL)	14.44 ± 3.76	13.73 ± 3.54	18.87 ± 4.67	<0.001
Uric acid (mg/dL)	4.68 ± 1.25	4.38 ± 1.13	5.96 ± 1.59	<0.001
Creatinine (mg/dL)	0.80 ± 0.17	0.71 ± 0.11	1.22 ± 0.20	<0.001

* Significant differences in the characteristics between the cases and controls were determined by the Student's *t*-test.

2.4. Genotyping

The genotype data were provided by the Center for Genome Science, Korea National Institute of Health. A total of 58,700 genomic DNA samples isolated from peripheral blood were genotyped using the Affymetrix Axiom® Array (Affymetrix, Santa Clara, CA, USA). The genotype data were confirmed using the Korean-Chip (K-CHIP, Seoul, Korea) acquired by the K-CHIP consortium. Detailed information on Korean chips has been described previously [19]. Samples with less than 96%–99% genotyping accuracy, excessive heterozygosity, or sex inconsistency were excluded. Markers with missing genotype rates of <95%, a minor allele frequency of <1%, and Hardy-Weinberg equilibrium *p*-values of <1 × 10^{-6} were also excluded to control the quality of the genotyping results. After quality control, imputation analysis was performed using the 1000 genome phase 3 dataset (reference panel) in IMPUTE v2 software. A total of 8,056,211 SNPs were included in the study. This study selected SNPs significantly related to kidney disease in the *MANBA* gene. The location of the SNPs was identified using National Center for Biotechnology Information (NCBI) Human Genome Build 37 (hg19).

2.5. Statistical Analysis

Statistical analyses were conducted with PLINK version 1.90 beta (https://www.cog-genomics.org/plink2 (accessed on 13 April 2021)) [20] and PASW Statistics version 18.0 (SPSS Inc. Chicago, IL, USA). A total of 58,700 participants were classified as CKD cases and controls according to KDIGO guidelines, and logistic regression analysis was performed to calculate the odds ratios (ORs) and 95% confidence intervals (95% CIs). Linear regression analysis was used to analyze the association between BUN, uric acid, and creatinine levels, and eGFR, which are traits related to kidney function, and *MANBA* genes. Logistic and linear regression analyses were performed based on the additive genetic model after age and gender adjustments. The significance threshold ($p < 6.76 \times 10^{-4}$) was adjusted through Bonferroni correction. Regional plots were created using the LocusZoom program (http://locuszoom.org/ (accessed on 13 April 2021)). A conditional analysis was performed to identify secondary association signals. The publicly available Human Kidney eQTL database (http://susztaklab.com/eqtl (accessed on 13 April 2021)) was used to determine whether the variants significantly associated with CKD affected the expression level of the *MANBA* gene.

3. Results

3.1. Participant Characteristics

The clinical characteristics of the 58,700 participants (20,293 men, and 38,407 women) in this study are shown in Table 1. To analyze the association between CKD and variants in the *MANBA* gene, CKD was divided into cases and controls using the eGFR. BUN, uric acid, and creatinine levels, which are related to kidney function, were increased in the case group compared to the control group. The comparison between the CKD cases and controls using the Student's *t*-test showed that all characteristics except BMI were significantly different.

3.2. Association Analysis of the MANBA Gene Variants with CKD and Kidney Function-Related Traits

The present study investigated the association of 229 SNPs in the *MANBA* gene with CKD and kidney function-related traits such as eGFR, BUN, creatinine, and uric acid levels. Before analyzing the association, we selected SNPs that tagged other SNPs with $r^2 < 0.8$ to show the independently related SNPs of the *MANBA* gene. As a result, 74 of 229 SNPs were identified, and finally, 20 SNPs achieved significance of $p < 6.76 \times 10^{-4}$ (Bonferroni threshold: $p = 0.05/74$) for association with CKD or kidney function-related traits in the HEXA cohort (Table 2). Most of the 20 SNPs are significantly associated with kidney function-related traits such as eGFR, creatinine and uric acid levels. However, there was no association between SNPs in the *MANBA* gene and BUN levels. The odds ratio of CKD was 0.87 (95% CI 0.80–0.95, $p = 2.61 \times 10^{-3}$) for the rs4496586 which showed the highest association with CKD. The linear regression analysis showed a significant association of rs4496586 with eGFR ($\beta = 0.565$, $p = 1.43 \times 10^{-9}$), creatinine ($\beta = -0.004$, $p = 1.04 \times 10^{-8}$), and uric acid levels ($\beta = -0.019$, $p = 1.83 \times 10^{-3}$). These results were consistent with the reduced risk of CKD in patients carrying the minor C allele of rs4496586.

Additionally, the Human Kidney eQTL database was used to analyze the association between the rs4496586 genotype, which had a high significance for CKD, and the *MANBA* gene expression in the renal tubules and glomeruli. The expression level of the *MANBA* gene was significantly increased in patients with the minor C allele of rs4496586 in the renal tubules and glomeruli (Figure 1).

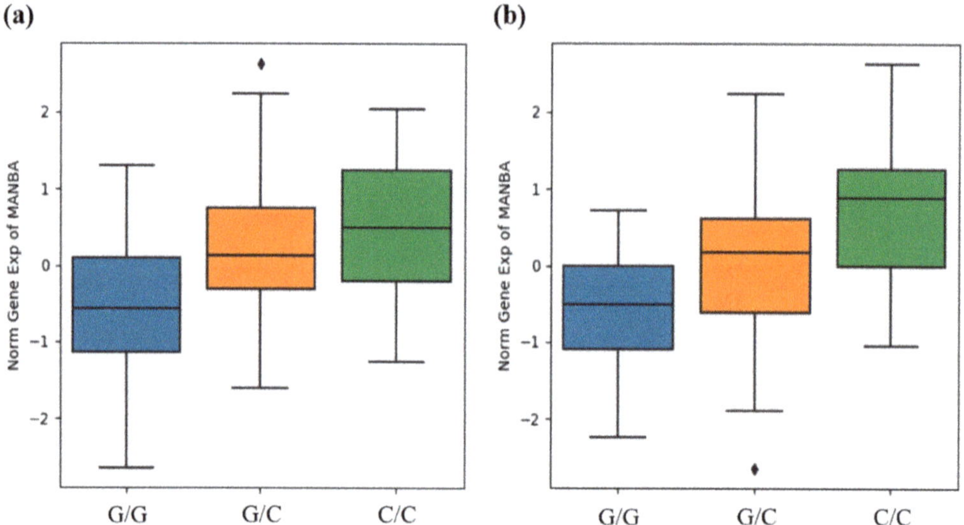

Figure 1. Rs4496586 genotype and *MANBA* gene expression association in human kidney tubules (**a**) and glomeruli (**b**). The data are from the Human Kidney eQTL database (http://susztaklab.com/eqtl (accessed on 13 April 2021)). *MANBA* gene expression for the rs4496586 genotype in tubules ($\beta = 0.636$, $p = 1.44 \times 10^{-6}$) and glomeruli ($\beta = 0.609$, $p = 2.49 \times 10^{-6}$) was confirmed and statistically significant. *p*-value was calculated by linear regression. Center lines show medians, box limits indicate 25th and 75th percentiles, and whiskers extend to the 5th and 95th percentiles, outliers are represented by diamonds.

Table 2. Results of association analysis of the SNPs in the *MANBA* gene with chronic kidney disease and kidney function-related traits ($r^2 < 0.8$).

No.	SNP	Minor Allele	MAF	Function	CKD		eGFR		Creatinine		Uric Acid	
					OR (95%CI)	*p*-Value	$\beta \pm$ S.E	*p*-Value	$\beta \pm$ S.E	*p*-Value	$\beta \pm$ S.E	*p*-Value
1	rs4496586	C	0.488	Intron	0.87 (0.80–0.95)	2.61×10^{-3}	0.565 ± 0.093	1.43×10^{-9}	-0.0042 ± 0.00073	1.04×10^{-8}	-0.019 ± 0.006	1.83×10^{-3}
2	rs223497	C	0.450	Intron	0.88 (0.81–0.97)	6.10×10^{-3}	0.406 ± 0.094	1.55×10^{-5}	-0.0034 ± 0.00073	3.91×10^{-6}	-0.013 ± 0.006	0.037
3	rs223489 *	G	0.357	Intron	0.88 (0.80–0.97)	7.08×10^{-3}	0.599 ± 0.098	9.42×10^{-10}	-0.0047 ± 0.00076	6.72×10^{-10}	-0.017 ± 0.006	8.61×10^{-3}
4	rs34768739	GA	0.455	Intron	0.89 (0.81–0.97)	7.65×10^{-3}	0.586 ± 0.094	3.86×10^{-10}	-0.0042 ± 0.00073	9.15×10^{-9}	-0.018 ± 0.006	3.18×10^{-3}
5	rs34642884	GC	0.471	Intron	0.89 (0.82–0.97)	9.30×10^{-3}	0.535 ± 0.093	1.05×10^{-8}	-0.0039 ± 0.00073	6.87×10^{-8}	-0.019 ± 0.006	2.07×10^{-3}
6	rs1054037	T	0.472	Intron	0.89 (0.82–0.98)	0.012	0.530 ± 0.093	1.41×10^{-8}	-0.0039 ± 0.00073	8.63×10^{-8}	-0.019 ± 0.006	2.36×10^{-3}
7	rs6847587 *	G	0.455	Intron	0.90 (0.82–0.98)	0.014	0.569 ± 0.094	1.22×10^{-9}	-0.0041 ± 0.00073	2.46×10^{-8}	-0.018 ± 0.006	3.13×10^{-3}
8	rs227361 *	C	0.451	Intron	0.90 (0.82–0.98)	0.015	0.571 ± 0.094	1.13×10^{-9}	-0.0041 ± 0.00073	2.62×10^{-8}	-0.019 ± 0.006	2.15×10^{-3}
9	rs228611 *	G	0.468	Intron	0.90 (0.82–0.98)	0.016	0.483 ± 0.094	2.51×10^{-7}	-0.0035 ± 0.00073	1.85×10^{-6}	-0.016 ± 0.006	9.43×10^{-3}
10	rs147730991	T	0.014	Intron	1.44 (1.05–1.98)	0.022	-0.396 ± 0.394	0.315	0.0042 ± 0.00307	0.175	-0.020 ± 0.026	0.429
11	rs36126232	CT	0.402	Intron	0.90 (0.82–0.99)	0.023	0.401 ± 0.096	2.76×10^{-5}	-0.0036 ± 0.00075	1.92×10^{-6}	-0.016 ± 0.006	0.012
12	rs223498	C	0.453	Intron	0.91 (0.83–0.99)	0.034	0.389 ± 0.094	3.49×10^{-5}	-0.0030 ± 0.00073	3.14×10^{-5}	-0.010 ± 0.006	0.111
13	rs223487	C	0.254	Intron	0.90 (0.81–0.99)	0.035	0.616 ± 0.107	9.73×10^{-9}	-0.0045 ± 0.00084	7.10×10^{-8}	-0.017 ± 0.007	0.016
14	rs11097790	C	0.384	Intron	0.91 (0.83–0.99)	0.038	0.432 ± 0.097	7.66×10^{-6}	-0.0029 ± 0.00075	9.06×10^{-5}	-0.012 ± 0.006	0.055
15	rs12650217	C	0.4637	Intron	0.92 (0.84–1.00)	0.054	0.355 ± 0.094	1.50×10^{-4}	-0.0028 ± 0.00073	1.46×10^{-4}	-0.0099 ± 0.0061	0.106
16	rs78905355	C	0.01218	Intron	0.69 (0.46–1.05)	0.084	1.586 ± 0.423	1.80×10^{-4}	-0.0123 ± 0.00330	1.92×10^{-4}	-0.0220 ± 0.0277	0.427
17	rs11454438	CA	0.4807	Intron	1.08 (0.99–1.17)	0.102	-0.400 ± 0.094	1.92×10^{-5}	0.0029 ± 0.00073	6.02×10^{-5}	0.0123 ± 0.0061	0.044
18	rs223503	G	0.2466	Intron	0.93 (0.84–1.03)	0.185	0.477 ± 0.108	1.08×10^{-5}	-0.0032 ± 0.00084	1.36×10^{-4}	-0.0099 ± 0.0071	0.164
19	rs170563 *	C	0.2613	Intron	0.94 (0.85–1.04)	0.213	0.480 ± 0.107	6.92×10^{-6}	-0.0033 ± 0.00083	7.33×10^{-5}	-0.0110 ± 0.0070	0.116
20	rs227374	A	0.2572	Intron	0.95 (0.86–1.05)	0.321	0.459 ± 0.107	1.86×10^{-5}	-0.0032 ± 0.00083	1.52×10^{-4}	-0.0108 ± 0.0070	0.123

SNP, single nucleotide polymorphism; MAF, minor allele frequency; CKD, chronic kidney disease; eGFR, estimated glomerular filtration rate; β, regression coefficient; S.E, standard error; OR, odds ratio; CI, confidence interval. eGFR, creatinine, and uric acid levels used in the linear regression were adjusted for age and gender. Odds ratios were calculated after adjusting for age and gender. Both logistic and linear regressions were conducted using an additive model. * SNPs replicated in the kidney-related results from other studies.

The SNPs of the *MANBA* gene, which were significant in the association analysis for CKD and kidney function-related traits, were shown using LocusZoom (http://csg.sph.umich.edu/locuszoom/ (accessed on 13 April 2021)), confirming the regional plots (Figure 2, Supplementary Figure S1). As a result, this study detected two independent signals through the regional plots. These SNPs of signal 1 (rs4496586) and signal 2 (rs223489) showed significantly high levels of CKD ($p = 2.61 \times 10^{-3}$ and $p = 7.08 \times 10^{-3}$), eGFR ($p = 1.43 \times 10^{-9}$ and $p = 9.42 \times 10^{-10}$), creatinine ($p = 1.04 \times 10^{-8}$ and $p = 6.72 \times 10^{-10}$), and uric acid ($p = 1.83 \times 10^{-3}$ and $p = 8.61 \times 10^{-3}$). To clearly identify secondary signals, this study performed conditional analyses, including the most significant SNPs of CKD, eGFR, creatinine, and uric acid in a stepwise manner. eGFR and creatinine found additional independent signals at the *MANBA* locus after conditioning for the most significant variants. In contrast, the conditional analyses for CKD and uric acid detected no conditionally independent signals (Supplementary Table S1).

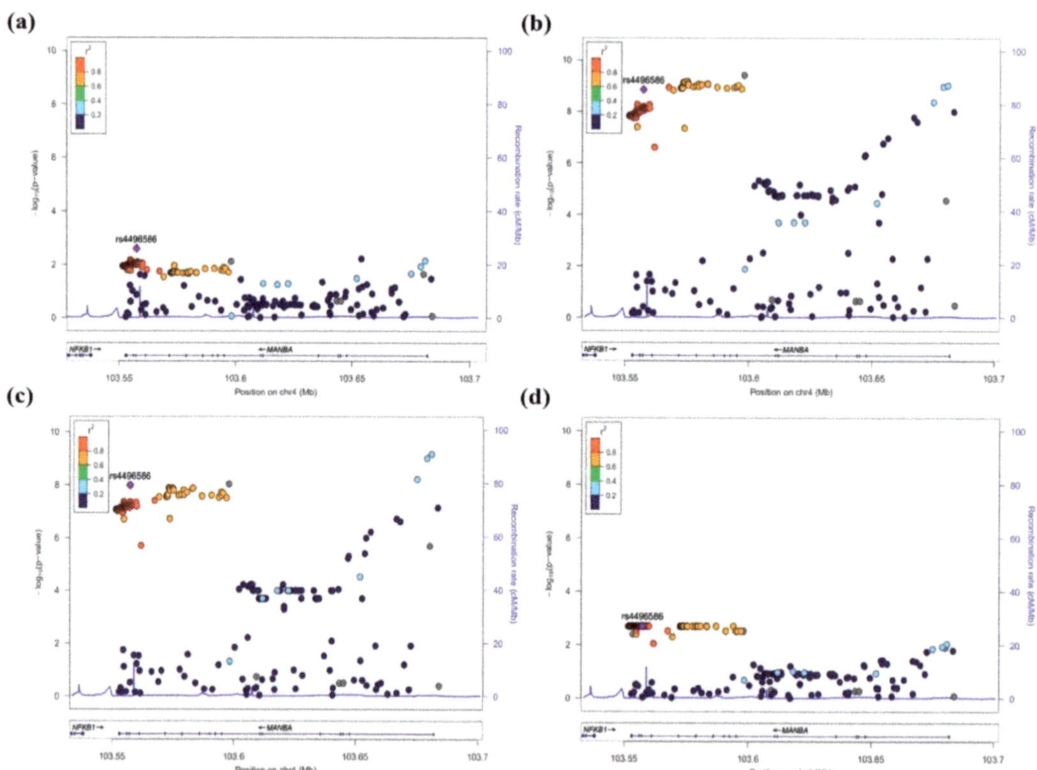

Figure 2. Regional plots of the association results of *MANBA* SNPs with CKD (**a**) and kidney function-related traits such as eGFR (**b**), creatinine (**c**), and uric acid levels (**d**) in the HEXA cohort. The statistical significance ($-\log_{10}$ *p*-value) of the analyzed SNPs is plotted. The colors indicate the linkage disequilibrium (r^2) between rs4496586 and the remaining SNPs. The genetic recombination rates are shown on the right *y*-axis. The plots were generated by the LocusZoom program (http://csg.sph.umich.edu/locuszoom (accessed on 13 April 2021)).

4. Discussion

CKD is a complex disease caused by impaired kidney function [21]. In addition, CKD is associated with serious complications such as cardiovascular disease, hyperlipidemia, hypertension, and metabolic bone disease [22–24]. The prevalence of CKD reported in Korea is lower than that confirmed in other countries [3,25]. However, the early diagnosis and treatment of CKD are important in preventing various complications [26]. This study performed an association analysis of variants in the *MANBA* gene with CKD and kidney function-related traits in a Korean cohort. The study found that 11 SNPs were not only statistically significant in CKD but were also significantly associated with kidney function-related traits such as creatinine and uric acid levels (Table 2). A previous study reported that the minor A allele of rs228611 was significantly associated with the eGFR ($\beta = -0.0056$, $p = 3.58 \times 10^{-12}$) [27]. Similar to previous results, this study showed that rs228611 with a minor G allele was associated with increases in the eGFR ($\beta = 0.483$, $p = 2.51 \times 10^{-7}$) (Supplementary Table S2). Another study identified rs223489 as a candidate eGFR variant (Effect allele = A, $\beta = -0.0027$, $p = 2.6 \times 10^{-17}$) by analyzing the human ortholog in the GWAS summary statistics (eGFR) for genes associated with abnormal kidney morphology in mice [8]. Therefore, it was suggested that rs223489, which was not previously identified as important, significantly affected the eGFR. Consistent with the previous study, our results also replicated the statistically significant association of rs223489 with the eGFR (Effect allele = G, $\beta = 0.599$, $p = 2.6 \times 10^{-10}$) in Koreans. In addition, rs223489 was detected as an important signal in CKD, eGFR, creatinine, and uric acid through the regional plots (Supplementary Figure S1).

Previous studies have reported a significant association between the eGFR and genetic variants in chromosome 4 through GWAS [9,28,29]. In addition, *NFKB1* was suggested as a target gene for kidney function [27]. Unlike previous studies, Ko et al. identified candidate genes for CKD through an integrative analysis that combined CKD-associated variants and kidney eQTL results [12]. The expression of *NFKB1* gene did not show any significance in the integrative analysis, and the *MANBA* gene, which showed statistically significant co-localization between CKD-associated variants and kidney eQTL results, was proposed as a potential target gene for the GWAS variant related to kidney function. Furthermore, the study revealed that kidney function was impaired when *MANBA* gene expression was suppressed in zebrafish. Thus, increasing the expression of the *MANBA* gene might be a potential new means to treat kidney dysfunction.

Qiu et al. suggested that many diseases are cell-type-specific, rather than organ-specific [11]. Thus, they performed eQTL analysis of the renal tubules and glomeruli and identified genes that may cause kidney disease, including the *MANBA* gene. In the present study rs4496586 had the highest significance in the association analysis between CKD and variants in the *MANBA* gene. Therefore, this study analyzed the association between the rs4496586 genotype and *MANBA* gene expression in the renal tubules and glomeruli. Our results showed that the expression level of the *MANBA* gene was significantly higher in samples with a minor allele of rs4496586 (Figure 1). In brief, rs4496586 increased the expression of the *MANBA* gene and decreased the risk of CKD in patients possessing a minor allele (C). These results are consistent with those of a previous study demonstrating the association between the expression of the *MANBA* gene and renal function in vivo [12].

Interestingly, a recent study conducted mechanistic experiments for the rs6847587 variant significantly associated with *MANBA* gene expression in renal tubules using mice and cells [15]. The results showed that a reduction in *MANBA* gene expression affected not only the structure and function of lysosomes, but also the endocytosis and autophagy pathways in vivo, leading to tubular damage, inflammation activation, and fibrosis. However, the study group mentioned the limitation of not using CKD-related variants. Our results from analyzing the association between CKD and *MANBA* gene variants showed that rs6847587 significantly reduced the risk of CKD in participants with the minor allele. The presence of the rs6847587 variant with the minor allele was associated with increased *MANBA* gene expression. Thus, these results are consistent with a previous study demonstrating the association of the *MANBA* gene in the development of kidney disease.

In summary, this study focused on the *MANBA* gene and confirmed the association of genetic variants with CKD and kidney function-related traits such as eGFR, BUN, creatinine, and uric acid levels based on KoGES. The *MANBA* gene variants showed significant associations with CKD, consistent with a recent study demonstrating that

MANBA gene variants were related to kidney function through an integrative analysis of eQTL and CKD-related GWAS results. Moreover, this study confirmed *MANBA* gene expression according to genotypes through eQTL analysis and demonstrated that the *MANBA* gene variants affecting renal tubules and glomeruli were significantly related to CKD. However, in vivo studies are needed to determine the direct effects of *MANBA* gene variants, which are highly correlated with CKD, on kidney function. Future studies also need to evaluate the association between CKD and *MANBA* gene variants through population-specific analysis.

Supplementary Materials: The following are available online at https://www.mdpi.com/article/10.3390/jcm10112255/s1, Table S1: Conditional analyses of multiple SNPs at the *MANBA* locus. Table S2: Results of an association analysis of the SNPs in the *MANBA* gene with chronic kidney disease and kidney function-related traits. Figure S1: Regional plots of the association results of *MANBA* SNPs with CKD (a) and kidney function-related traits such as eGFR (b), creatinine (c), and uric acid levels (d) in the HEXA cohort. The statistical significance ($-\log_{10}$ p-value) of the analyzed SNPs is plotted. The colors indicate the linkage disequilibrium (r^2) between rs223489 and the remaining SNPs. The genetic recombination rates are shown on the right y-axis. The plots were generated by the LocusZoom program (http://csg.sph.umich.edu/locuszoom/ (accessed on 13 April 2021)).

Author Contributions: Conceptualization, H.-R.K. and H.-S.J.; methodology, H.-S.J.; software, H.-R.K.; validation, H.-R.K., H.-S.J. and Y.-B.E.; investigation, H.-R.K.; resources, Y.-B.E.; data curation, H.-R.K. and H.-S.J.; writing—original draft preparation, H.-R.K. and H.-S.J.; writing— review and editing, H.-S.J. and Y.-B.E.; supervision, H.-S.J. and Y.-B.E.; project administration, Y.-B.E.; funding acquisition, Y.-B.E. All authors have read and agreed to the published version of the manuscript.

Funding: This research was supported by the Soonchunhyang University research fund and a National Research Foundation of Korea (NRF) grant funded by the Korean government (MSIT) [NRF-2020R1F1A1071977].

Institutional Review Board Statement: This study was approved by the Institutional Review Board of the Korea Disease Control and Prevention Agency (KDCA, KBN-2021-003 (26 January 2021)) and Soonchunhyang University (202012-BR-086-01 (15 December 2020)).

Informed Consent Statement: Informed consent was obtained from all subjects involved in the study.

Data Availability Statement: The data presented in this study are available on request from the corresponding author. The data are not publicly available due to ethnical concerns.

Conflicts of Interest: The authors declare no conflict of interest.

References

1. Levey, A.S.; Atkins, R.; Coresh, J.; Cohen, E.P.; Collins, A.J.; Eckardt, K.U.; Nahas, M.E.; Jaber, B.L.; Jadoul, M.; Levin, A.; et al. Chronic kidney disease as a global public health problem: Approaches and initiatives—A position statement from Kidney Disease Improving Global Outcomes. *Kidney Int.* **2007**, *72*, 247–259. [CrossRef]
2. Tonelli, M.; Wiebe, N.; Culleton, B.; House, A.; Rabbat, C.; Fok, M.; McAlister, F.; Garg, A.X. Chronic kidney disease and mortality risk: A systematic review. *J. Am. Soc. Nephrol.* **2006**, *17*, 2034–2047. [CrossRef]
3. Carney, E.F. The impact of chronic kidney disease on global health. *Nat. Rev. Nephrol.* **2020**, *16*, 251. [CrossRef]
4. Parmar, M.S. Chronic renal disease. *BMJ* **2002**, *325*, 85–90. [CrossRef]
5. MacCluer, J.W.; Scavini, M.; Shah, V.O.; Cole, S.A.; Laston, S.L.; Voruganti, V.S.; Paine, S.S.; Eaton, A.J.; Comuzzie, A.G.; Tentori, F.; et al. Heritability of measures of kidney disease among zuni Indians: The Zuni Kidney Project. *Am. J. Kidney Dis.* **2010**, *56*, 289–302. [CrossRef]
6. Langefeld, C.D.; Beck, S.R.; Bowden, D.W.; Rich, S.S.; Wagenknecht, L.E.; Freedman, B.I. Heritability of GFR and albuminuria in Caucasians with type 2 diabetes mellitus. *Am. J. Kidney Dis.* **2004**, *43*, 796–800. [CrossRef]
7. Smyth, L.J.; Duffy, S.; Maxwell, A.P.; McKnight, A.J. Genetic and epigenetic factors influencing chronic kidney disease. *Am. J. Physiol.-Ren.* **2014**, *307*, F757–F776. [CrossRef]
8. Wuttke, M.; Li, Y.; Li, M.; Sieber, K.B.; Feitosa, M.F.; Gorski, M.; Tin, A.; Wang, L.; Chu, A.Y.; Hoppmann, A.; et al. A catalog of genetic loci associated with kidney function from analyses of a million individuals. *Nat. Genet.* **2019**, *51*, 957–972. [CrossRef]
9. Okada, Y.; Sim, X.; Go, M.J.; Wu, J.Y.; Gu, D.; Takeuchi, F.; Takahashi, A.; Maeda, S.; Tsunoda, T.; Chen, P.; et al. Meta-analysis identifies multiple loci associated with kidney function-related traits in east Asian populations. *Nat. Genet.* **2012**, *44*, 904–909. [CrossRef]

Morris, A.P.; Le, T.H.; Wu, H.; Akbarov, A.; van der Most, P.J.; Hemani, G.; Smith, G.D.; Mahajan, A.; Gaulton, K.J.; Nadkarni, G.N.; et al. Trans-ethnic kidney function association study reveals putative causal genes and effects on kidney-specific disease aetiologies. *Nat. Commun.* **2019**, *10*, 29. [CrossRef] [PubMed]

Qiu, C.; Huang, S.; Park, J.; Park, Y.; Ko, Y.A.; Seasock, M.J.; Bryer, J.S.; Xu, X.X.; Song, W.C.; Palmer, M.; et al. Renal compartment-specific genetic variation analyses identify new pathways in chronic kidney disease. *Nat. Med.* **2018**, *24*, 1721–1731. [CrossRef] [PubMed]

Ko, Y.A.; Yi, H.G.; Qiu, C.X.; Huang, S.Z.; Park, J.; Ledo, N.; Kottgen, A.; Li, H.Z.; Rader, D.J.; Pack, M.A.; et al. Genetic-Variation-Driven Gene-Expression Changes Highlight Genes with Important Functions for Kidney Disease. *Am. J. Hum. Genet.* **2017**, *100*, 940–953. [CrossRef] [PubMed]

Howard, M.F.; Murakami, Y.; Pagnamenta, A.T.; Daumer-Haas, C.; Fischer, B.; Hecht, J.; Keays, D.A.; Knight, S.J.; Kolsch, U.; Kruger, U.; et al. Mutations in *PGAP3* impair GPI-anchor maturation, causing a subtype of hyperphosphatasia with mental retardation. *Am. J. Hum. Genet.* **2014**, *94*, 278–287. [CrossRef]

Araki, T.; Hayashi, M.; Nakanishi, K.; Morishima, N.; Saruta, T. Caspase-9 takes part in programmed cell death in developing mouse kidney. *Nephron Exp. Nephrol.* **2003**, *93*, e117–e124. [CrossRef]

Gu, X.; Yang, H.; Sheng, X.; Ko, Y.A.; Qiu, C.; Park, J.; Huang, S.; Kember, R.; Judy, R.L.; Park, J.; et al. Kidney disease genetic risk variants alter lysosomal beta-mannosidase (*MANBA*) expression and disease severity. *Sci. Transl. Med.* **2021**, *13*, eaaz1458. [CrossRef]

Bedilu, R.; Nummy, K.A.; Cooper, A.; Wevers, R.; Smeitink, J.; Kleijer, W.J.; Friderici, K.H. Variable clinical presentation of lysosomal beta-mannosidosis in patients with null mutations. *Mol. Genet. Metab.* **2002**, *77*, 282–290. [CrossRef]

Sabourdy, F.; Labauge, P.; Stensland, H.M.; Nieto, M.; Garces, V.L.; Renard, D.; Castelnovo, G.; de Champfleur, N.; Levade, T. A *MANBA* mutation resulting in residual beta-mannosidase activity associated with severe leukoencephalopathy: A possible pseudodeficiency variant. *BMC Med. Genet.* **2009**, *10*, 84. [CrossRef]

Kim, Y.; Han, B.G.; KoGES Group. Cohort Profile: The Korean Genome and Epidemiology Study (KoGES) Consortium. *Int. J. Epidemiol.* **2017**, *46*, 1350. [CrossRef] [PubMed]

Moon, S.; Kim, Y.J.; Han, S.; Hwang, M.Y.; Shin, D.M.; Park, M.Y.; Lu, Y.; Yoon, K.; Jang, H.M.; Kim, Y.K.; et al. The Korea Biobank Array: Design and identification of coding variants associated with blood biochemical traits. *Sci. Rep.* **2019**, *9*, 1382. [CrossRef]

Purcell, S.; Neale, B.; Todd-Brown, K.; Thomas, L.; Ferreira, M.A.; Bender, D.; Maller, J.; Sklar, P.; de Bakker, P.I.; Daly, M.J.; et al. PLINK: A tool set for whole-genome association and population-based linkage analyses. *Am. J. Hum. Genet.* **2007**, *81*, 559–575. [CrossRef]

Romagnani, P.; Remuzzi, G.; Glassock, R.; Levin, A.; Jager, K.J.; Tonelli, M.; Massy, Z.; Wanner, C.; Anders, H.J. Chronic kidney disease. *Nat. Rev. Dis. Primers* **2017**, *3*, 17088. [CrossRef]

Thomas, R.; Kanso, A.; Sedor, J.R. Chronic kidney disease and its complications. *Prim. Care* **2008**, *35*, 329–344. [CrossRef] [PubMed]

Go, A.S.; Chertow, G.M.; Fan, D.; McCulloch, C.E.; Hsu, C.Y. Chronic kidney disease and the risks of death, cardiovascular events, and hospitalization. *N. Engl. J. Med.* **2004**, *351*, 1296–1305. [CrossRef] [PubMed]

National Kidney, F. K/DOQI clinical practice guidelines for bone metabolism and disease in chronic kidney disease. *Am. J. Kidney Dis.* **2003**, *42*, S1–S201. [CrossRef]

Bikbov, B.; Purcell, C.; Levey, A.S.; Smith, M.; Abdoli, A.; Abebe, M.; Adebayo, O.M.; Afarideh, M.; Agarwal, S.K.; Agudelo-Botero, M.; et al. Global, regional, and national burden of chronic kidney disease, 1990–2017: A systematic analysis for the Global Burden of Disease Study 2017. *Lancet* **2020**, *395*, 709–733. [CrossRef]

Fraser, S.D.; Blakeman, T. Chronic kidney disease: Identification and management in primary care. *Pragmat. Obs. Res.* **2016**, *7*, 21–32. [CrossRef]

Pattaro, C.; Teumer, A.; Gorski, M.; Chu, A.Y.; Li, M.; Mijatovic, V.; Garnaas, M.; Tin, A.; Sorice, R.; Li, Y.; et al. Genetic associations at 53 loci highlight cell types and biological pathways relevant for kidney function. *Nat. Commun.* **2016**, *7*, 10023. [CrossRef]

Pattaro, C.; Kottgen, A.; Teumer, A.; Garnaas, M.; Boger, C.A.; Fuchsberger, C.; Olden, M.; Chen, M.H.; Tin, A.; Taliun, D.; et al. Genome-wide association and functional follow-up reveals new loci for kidney function. *PLoS Genet.* **2012**, *8*, e1002584. [CrossRef]

Kottgen, A.; Glazer, N.L.; Dehghan, A.; Hwang, S.J.; Katz, R.; Li, M.; Yang, Q.; Gudnason, V.; Launer, L.J.; Harris, T.B.; et al. Multiple loci associated with indices of renal function and chronic kidney disease. *Nat. Genet.* **2009**, *41*, 712–717. [CrossRef]

MDPI

St. Alban-Anlage 66

4052 Basel

Switzerland

Tel. +41 61 683 77 34

Fax +41 61 302 89 18

www.mdpi.com

Journal of Clinical Medicine Editorial Office

E-mail: jcm@mdpi.com

www.mdpi.com/journal/jcm